Earth
at Hand

Acknowledgements

We wish to gratefully acknowledge the sustained effort of our spouses, Angie Callister and Richard Stroud. We also want to thank the Publications Staff of NSTA, Phyllis R. Marcuccio, Shirley Watt Ireton, Andrew Saindon, and Daniel T. Shannon for providing us with their continuous support and words of guidance throughout this project.

The greatest credit goes to the staffs of the NSTA journals—*Science & Children, Science Scope, The Science Teacher*, and the *Journal of College Science Teaching*. Their insight and talent at checking for accuracy and ensuring clarity of phrasing has enabled us to assemble this valuable collection for teachers.

Earth at Hand

A collection of articles
from NSTA's journals

Collected by
SHARON M. STROUD
JEFFREY C. CALLISTER

JOURNAL OF
COLLEGE
SCIENCE
TEACHING

The
Science
Teacher

SCIENCE
Scope

SCIENCE
&CHILDREN

National Science Teachers Association

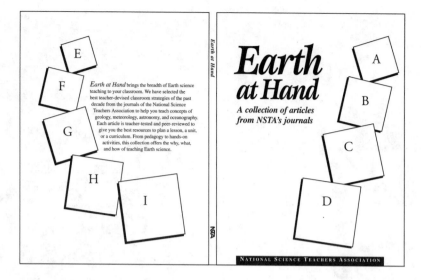

Earth at Hand

A collection of articles from NSTA's journals

NATIONAL SCIENCE TEACHERS ASSOCIATION

About the Cover

Background photo This image of a volcanic eruption on Io, one of Jupiter's satellites, was taken by Voyager I about half a million kilometers from the surface. The image is a special color reconstruction compiled from four pictures taken in ultraviolet, blue, green, and orange light. Data from this type of image provides information on the amount of gas and dust in the eruption and the size of the dust particles. Photo courtesy of NASA.

A. This outcrop of stone pillars are of Precambrian gneiss, a metamorphic rock. They are jutting up from a field on Lewis Island of the outer Hebrides off the western coast of Scotland. The gneiss on Lewis Island is often referred to as Lewissian gneiss. Photo courtesy of Dr. Dorothy Stout, professor of geology at Cypress College in California and past-president of the National Association of Geology Teachers.

B. These branched corals in Jamaica's Discovery Bay are classified as *Acropora palmata.* Photo courtesy of Christopher F. D'Elia of the Sea Grant College Program at the University of Maryland.

C. This bolt of lightening struck over Gaithersburg, Maryland. The thunderstorm came from the north with winds blowing to 40 knots. Photo courtesy of H. Michael Mogil.

D. Photo courtesy of Jim Green.

E. These corals are part of the Enewetak Atoll in the Marshall Islands of the South Pacific. Virtually all of the corals belong to the genera *Acropora* or *Pocillopora.* Photo courtesy of Christopher F. D'Elia of the Sea Grant College Program at the University of Maryland.

F. These cirrostratus clouds at sunrise foretell of rain later in the day. The prefix "cirro" indicates the base of these clouds is hovering above 6,000 meters, and "stratus" refers to their layered structure. Photo taken at Rockville, Maryland, courtesy of H. Michael Mogil.

G. This soccer ball-like nodule, called a Septanian nodule or a Muraki boulder, was weathered out of the sedimentary coastal rock of South Island in New Zealand. Photo courtesy of Dr. Dorothy Stout.

H. This color-enhanced image shows the varying density of the Sun's corona. The purple regions, representing highest coronal density, overlie sunspot regions on the solar surface. (Lowest coronal density is represented in yellow.) The largest spike extends beyond 1.5 million kilometers above the Sun's surface. This image was prepared from data supplied by NASA's Solar Maximum Mission Satellite. Photo courtesy of NASA.

I. Photo courtesy of National Science Resources Center.

Table of Contents

Chapter 2: Water, Weather, and the Environment   63

Chapter 3: The Earth in Space   105

Introduction

In recent years, the science community has witnessed a significant increase in public awareness of Earth science, and with good reason. No other science discipline has such a far reaching impact and few have undergone the variety of interesting and progressive changes that Earth science has, of late. The concept of a small "island" Earth, the first survey of the solar system, increased evidence of a structured universe, and complex environmental relationships have all forced this most interdisiplinary of subjects to the forefront of public attention.

In this explosion of knowledge, Earth science educators have built on foundations provided by many professional and educational organizations that have worked separately and in collaboration with NSTA to foster Earth science literacy. Achieving such literacy in the classroom requires a combination of approaches and techniques including investigative learning, minds-on approach, science-technology-society interactions, a global environmental view, and problem solving—all of which are reflected in this publication. Good activities, such as those in *Earth at Hand*, should be equally useful for leaders of inservice courses for teachers as well as directly in the classroom by teachers.

The editors spent many hours reading and selecting from hundreds of Earth science articles from NSTA's journals—*Journal of College Science Teaching, The Science Teacher, Science Scope,* and *Science and Children*—from January, 1982 to May, 1991. Their goal for this publication is to provide teachers with supplementary Earth science activities for a range of grade levels and subject matter.

As with all good teaching, these articles usually have many connections to many other areas of science teaching, and often include interdisciplinary connections as well. The editors have organized the articles by their primary focus, with the following internal order.

• Earth's Properties and Features: The age of the Earth and prehistoric life, Geologic processes, Minerals and natural resources, Soil, Ground water, Topography and geography.

• Water, Weather, and the Environment: Oceanography, Environmental concerns, Meteorology.

• The Earth in Space: Seeing stars, Our star, Our planet, Our solar system.

The interdisciplinary nature of Earth science could allow for several of these articles to be categorized under more than one topic, and a perusal of all articles may lead to innovative linking of topics. These articles were originally written for specific age ranges, mostly for grades 5–10. Within the limits of concept readiness, most articles may be adapted to enhance your teaching at any grade level.

Articles were selected that concentrate on activities with minimal or no background material. The articles involve classroom activities that can be used from elementary grades through the college level. Most activities use simple, readily obtainable materials and supplies and should require little, if any, further research. The editors feel that this publication will provide a valuable resource for investigative education and will enhance the fostering of an informed population in the Earth sciences, as well as science in general.

The second part of this publication is a bibliography which lists articles involving all aspects of Earth science education. The bibliography lists the Earth science articles from the four NSTA journals during the period of January, 1982 through May, 1991. It lists the many activity-based articles not included in this publication plus any other articles that were Earth Science related. This bibliography should be especially useful to individuals reviewing the literature in Earth science education.

Jeffrey C. Callister and Sharon M. Stroud

Earth's
Properties
and
Features

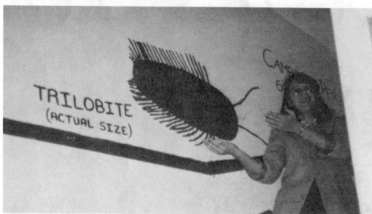

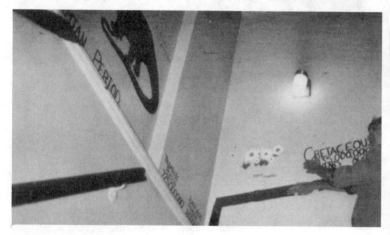

GIANT GEOLOGIC TIME

When facing twenty-six seventh graders struggling with the concept of geologic time, teachers need a clear and meaningful lab experience. I tried earth science resource books, but found few ideas on how to present vast time expanses in a way that thirteen-year-olds can comprehend. Many texts present the earth's history in clock or calendar form. Students can perceive time as linear but have difficulty with finite forms. I visited the Museum of Natural History in New York and saw their clear geologic time line. With this model in mind, I had my students construct a time line of their own using a metric scale and illustrations. I received approval and $50 for expenses from the school headmaster to undertake the time line project.

The size of the time line is important. It has to be long enough to illustrate 4.6 billion years. I teach in a crowded seven-story school and the only free wall space where a time line would fit was the back stairway. We made the 4.6 billion-year line 10 centimeters wide and 92 meters long. Each meter equals 50 million years. It starts on the first floor and ascends the walls of the stairwell to the sixth floor. Each geologic period is a different color, and a key to the time line is on the fourth floor.

Planning Ahead

It is important to give every student a meaningful job and the project a five-week deadline. I let students choose their own jobs from a list of opportunities and qualifications (Table 1). To make project implementation easier, the students researched and devised Time Line Specifications (Table 2) before they started painting.

Table 1. Student Labor Division

Title	Number of Openings	Job Description	Qualifications
Engineer	4	Measure and lay out time line	Good math and measuring skills
Technician	4	Draw line to engineer specifications	Neat and diligent
Painter	6	Paint line on wall	Neat and patient
Colorist	2	Choose and mix colors	Creative and artistic
Artist	4	Design and paint illustrations	Artistic
Letterer	2	Letter all captions	Good penmanship
Messenger	2	Deliver instructions and paint up and down stairs	Athletic
Foreman	2	Monitor all workers and report time line's progress	Good leadership and communication skills

Table 2. Time Line Specifications*

Geologic Time Period	Length (years)	Meters	Color	Illustrations
Precambrian	4.0 billion	80	Blue-gray	Fiery Earth Volcanoes
Cambrian	100 million	2	Blue	Trilobite
Ordovician	60 million	1.2	Dark green	Crinoid
Silurian	40 million	0.8	Yellow	Coelacanth
Devonian	50 million	1.0	Orange	
Carboniferous	80 million	1.6	Umber	Fern
Permian	45 million	0.9	Dark brown	
Triassic	45 million	0.9	Light green	Brontosaurus
Jurassic	45 million	0.9	Aqua	Flowering plants
Cretaceous	65 million	1.3	Green	
Tertiary	69 million	1.38	Purple	Saber-tooth tiger
Quaternary	1 million	0.002	Pink	Neanderthals

*In light of new research, many of the geologic time periods used are now approximations.

Working Together

Students divided themselves into two lab sections. Each section worked one 80-minute lab period per week for five weeks. The students worked in teams so they wouldn't all be working on the same part of the time line at one time. Engineers carefully measured the lengths of each geologic period while technicians drew double lines ten centimeters apart in pencil. Painters taped newspapers to the wall and painted the line. Artists painted huge algae, trilobites, coelacanths, dinosaurs, saber-toothed tigers, and Neanderthals. Each illustration was captioned by the letterers. The colorist created and matched paint colors using only primary colors and black and white. Though working in groups, students each had individual responsibilities, so they were able to spread out over the six floors and work without constant supervision.

The Finished Product

The line climbed up the stairway in a spiral that went over doors and around windows. The Precambrian Era, an ominous dark blue-gray, stretched from a fiery earth on the first floor 62 meters to the fifth floor. (Observers are always impressed by the length of time it took for the earth to cool enough to support life.) Each successive era was painted a brighter shade, culminating in a two-centimeter-wide bright pink Quaternary period.

The time line has been a valuable addition to our school. It is shown to visiting parents, and used by earth science and biology teachers to discuss evolution and geologic time. The students who constructed the time line not only learned a lot about geologic periods but also experienced a great deal of satisfaction in making a permanent contribution to the school.

JANET L. VILLAS
The Birch Wathen School
New York, New York

THE DINOSAURS CAME IN DECEMBER...
Putting Geologic Time Into Perspective

Classroom Calendar Combined with Geologic Timeline

											CENOZOIC
										PALEOZOIC	
										MESOZOIC	
— — — — — — — — — — — — PRECAMBRIAN — — — — — — — — — — — —											T O D A Y
Jan	Feb	Mar	Apr	May	June	July	Aug	Sep	Oct	Nov	Dec

Figure 1:

What is time? Time is of the essence. Time is on our side. Time flies when you're having fun. Time turns potentialities into actualities. Yet for most middle and junior high school students, the concept of time is very abstract and leads to difficulties in understanding the Earth's immense history.

Giving students a comprehensible frame of reference for the vastness of the Earth's past can tax the ingenuity of the most skillful teacher. One solution is to develop a 365-day calendar and a time line to be used as analogies of the Earth's 4.5 to 4.6 billion year history. This will give students a concrete cognitive tool for understanding the relative length of each geological era and the sequence of such important biological events as the appearance of plants, amphibians, birds, dinosaurs, and man.

The Creative Process

Have students brainstorm in groups to generate a list of events that are part of the Earth's past. Students will draw on their experiences to name topics such as dinosaurs, Pangaea, trilobites, and cave dwellers (the Fintstones don't count).

After students choose an event from the list, they spend time in the library doing research. Ask them to find the time in geologic history when their chosen event or subject made its mark. Later, they will use this information to place the event on the calendar and on the geologic time line.

Students choose their own method for displaying the collected data. Cartoons, cave etchings, posters, poems, stories, reports, collages, and bulletin boards are among endless possibilities for creativity and for integrating other disciplines and skills.

Calculations

After checking the accuracy of their historical information, students have to calculate where to place their events or subjects on a year-long calendar. Contacting businesses in the community will often result in enough contributions to provide each student (or student group) with a calendar.

Use calculators and the following conversion formula to find the analogous calendar date.

$$\text{Days} = \frac{\text{Event Age in Millions of Years}}{\text{Earth's Age in Millions of Years}} \times 365$$

A sample problem may follow this sequence: The first birds appeared 150 million years ago. The age of the Earth is estimated at 4.6 billion years.

4

DECEMBER

SUN	MON	TUES	WED	THURS	FRI	SAT
			1	**2**	**3**	
Mississippian Period Begins **4**	1. Appalachians Rise 2. Coal Forests Begin **5**	First Reptiles Appear in Fossil Record **6**	**7**	**8**	Permian Period Begins **9**	**10**
Mesozoic Era Begins **11**	**12**	1. Tyranasaurus Rex Dominates 2. First Mammals **13**	**14**	**15**	Jurassic Period Begins **16**	**17**
18	First Birds Appear in Fossil Record **19**	**20**	**21**	**22**	**23**	**24**
1. Dinosaurs Become Extinct 2. Rockies Rise **25**	First Elephants Appear **26**	**27**	**28**	**29**	**30**	1. First Humans 2. Ice Age Begins **31**

Figure 2:

Mathematically, these two numbers can be compared to a year of 365 days as follows:

$$\text{Days} = \frac{150,000,000}{4,600,000,000} \times 365$$

Students will find that Days = 11.9. Rounding this off to 12 days, students then subtract 12 days from the end of the year, December 31, which represents the present in this sequence. Thus, on this calendar, the first birds appeared December 19. Students will recognize that this event occurred, relatively speaking, in the not-so-distant geologic past.

The Classroom Calendar

After each student has used the equation to determine the equivalent days that represent the elapsed geologic time for each event chosen, construct a large class calendar. (See Figure 2.)

Make the daily calendar squares large enough for students to record their results. Display the calendar in a central location, within easy view of all students.

With the class display calendar waiting to be filled with information, students share their research creations with the class. After each student finishes his or her presentation, the class guesses when the event occurred as represented by the year-long calendar. (Each day represents about 12.6 million years.) Geologic time takes on a new perspective as students record their information on the appropriate daily squares.

Constructing a Geologic Time Line

A geologic time line of the events shown on a calendar can be designed to add another perspective to help students understand geologic time. The time scale for the wall of a classroom can be figured at 1 mm for every 1 million years. To represent our planet's age, I used 4.6 meters of adding machine tape. To record geological or biological events on the time line, subtract from the far right end labeled "Today."

Extending and Enriching

Students can construct a combination of the calendar and the time line to link both perspectives. To do this, students need to establish a length that represents the passage of one day on this time line. To find this figure, divide 4,600 (the time line's length in mm) by 365 (days per year). The result is 12.6. A calendar day, then, would be 12.6 mm long on the time line. To figure out how many mm to allot to a particular month, multiply 12.6 by the numbers of days in that month. (See Figure 1.)

Other enrichment activities include extending the time line into the future to show scientific predictions for such events as the burn-out of the sun or depletion of natural resources. Also, students can creatively illustrate or write about their own "future" time lines. This activity builds reasoning skills because the students are forced to base their predictions on information that is presently available.

By doing research and collaborating with classmates, students learn about geologic time and processes and gain a perspective not only of the Earth's past, but also of their own.

ARNOLD D. LINDAMAN
And
MARK MC CARTHY
North Scott Community School District
Eldridge, Iowa

Peachaceous

Late Lemonian

Early Lemonian

Limaceous

BLUEBERRY TRILOBITES

Betty Crocker
Edward L. Shaw, Jr.

Are you prospecting for a way to help students grasp geological concepts like horizontality, superposition, and stratification? You could take them to examine some nearby roadcuts or rock outcrops. Or, if a field trip is impractical, you could always fall back on photographs or clay models. But why not give your lesson a different flavor by opting for a unit on Jell-O geology? The materials are inexpensive and easy to work with, and geological concepts taught with the familiar dessert will make a big impression.

Trade Your Pickax for a Spoon

You'll need several packages of Jell-O (or another similar gelatin dessert) in assorted flavors, clear plastic cups for each student participating, an electric teakettle (or some other way of heating water to the boiling point), spoons, and the goodwill of the cafeteria personnel.

If you decide to investigate how geologists identify strata by fossils, you'll also want some pieces of fruit to stand in as your trilobites or ammonites—blueberries and sliced peaches or bananas work well. A supply of masking tape will allow students to identify their geological creations as their own. And the tops or bottoms of several heavy cardboard boxes—like the ones canned soft drinks come in—will make handy trays for collecting and transporting the models to and from the cafeteria refrigerator. (The cafeteria staff will be impressed by the thoughtfulness of this last measure.)

Jell-O Geology

Begin by introducing students to the law of superposition, which helps geologists establish a relative time sequence for the geological events they study. The law of superposition is used primarily to determine the relative ages of stratified rocks, or rocks that appear in layers (strata). It holds that before the topmost stratum in any sequence of

layered rocks could be laid down, the layer below it must already have been deposited. Thus, in any sequence of layered rocks, a given layer will be older than the layer above it and younger than the layer below.

Following the directions on the package of Jell-O, students can easily build models that demonstrate superposition. Have them dissolve one of the flavors in hot water, add cold water in the amount the package indicates, and spoon the cooled mixture into their plastic cups. Let this first layer set overnight, and repeat the procedure the next day with a different colored gelatin. Once students have several layers in place, they'll have internalized the idea that deposition requires a great deal of time (a concept that doesn't come across with conventional clay models), and they'll be able to date the different deposits in their own models.

The law of horizontality can also be illustrated with Jell-O strata. This law states that sediments are originally de-

AND PEACH STRATA

posited in a horizontal (or nearly horizontal) plane and parallel to the surface on which they are deposited. Thus, if a sequence of strata is not horizontal but rather tipped or tilted at an angle, scientists conclude that the event that produced the tilting must have occurred after the strata were laid down. (Exceptions to this rule do exist, but they are rare.) Some further experimenting with Jell-O (can students find a way to layer the gelatin on a slope? vertically?) will help solidify this concept.

Dating the Papaya Chunks

Stratified Jell-O can help explain other concepts, too. Scientists, for example, may be able to establish a relative time sequence for the creation of strata in a given area. But how can they correlate rocks in one area with those in another, especially if the areas are continents apart? One method is to compare the fossils found in the layers of rock they are studying. Layers containing the same kinds of fossils—that is, fossils of the same geological age—are presumed to be also of the same age.

A highly simplified version of this approach to dating can be illustrated by creating Jell-O strata that contain fruit "fossils." Make a master mold using a large glass bowl with fruit pieces of various kinds distributed in the different strata. (Establish a chronology that governs which Jell-O stratum will get blueberries and which peaches, but don't make the chronology explicit to your students. And, for simplicity's sake, put no more than two kinds of fossils in a layer.) Now, have students construct a chronology based on the fossil evidence in the mold. Did blueberries disappear after the Cherrycene Era? Were papaya chunks common only during the Lemonic Period?

When students have finished, provide them with a set of several smaller molds whose strata may or may not match the master mold but whose layers contain some of the same kinds of fossils and ask them to determine the relative age of the strata in these smaller molds.

Some Further Trifling

By the time you've finished this unit, your Jell-O may still wobble, but your students won't: they'll have a grasp of some basic geologic phenomena and will be more likely to remember what they've learned than if they'd simply discussed some clay models or looked at some photographs. And once they've consumed a few strata or polished off fossils from the Peachaceous Era, they may get some edible geology ideas of their own. (What about an English trifle, with some whipped cream to show glacial action?)

Betty Crocker is a doctoral student in the Department of Science Education, University of Georgia, Athens; Edward L. Shaw, Jr., is an assistant professor of education, Southeastern Louisiana University, Hammond. Artwork by Johanna Vogelsang.

Inferring About Dinosaurs

— Cynthia Szymanski Sunal —

One hundred twenty million years ago, armored *Ankylosaurus* and three-horned, wide-collared *Triceratops* walked the Earth, together with savage *Tyrannosaurus rex* and gigantic *Brontosaurus*. But did the creatures walk with long, slow strides or with rapid, birdlike movements? Did they have constant, warm body temperatures, or did their temperatures change with their surroundings? And did the plant-eaters among them browse high in treetops, or did they graze along riverbanks and bottoms?

Scientists don't have the answers to all these questions, but they have used fossil bones, footprints, and knowledge of the conditions under which these were made to construct models—tentative schemes based on inferences drawn from the evidence available. And while you probably don't have a skeleton, or even a dinosaur footprint, to work with in your class, your students can learn about both dinosaurs and model making as they investigate *what* paleontologists know about dinosaurs and *how* they know it.

Building Models

When it comes to studying dinosaurs, what we know is not as important as how we know it. Much science consists of building models based on limited evi-

found, and reassembling them involves inferences that may or may not be correct. So bones can go back in the wrong places. Or some may be missing, leaving scientists without an important piece of the puzzle until a similar skeleton is found with the missing pieces in place. Paleontologists have their work cut out for them.

The Skeleton in the Classroom

Introduce your unit on dinosaur studies by asking if anyone in the class passed a dinosaur on the way to school. Disregard unbelieving looks and giggles and ask, "Why not?" Encourage students to discuss whatever they know about dinosaurs and summarize their observations on the chalkboard.

Go on to give students a taste of how paleontologists work by having them reconstruct a dinosaur skeleton. To prepare their cache of bones, photocopy a diagram of a dinosaur (a side view will work best), and cut bones or groups of bones apart. Rearrange the parts by scattering them over another sheet of paper, and photocopy the rearranged bones. (Young children may have difficulty working with tiny scraps of paper so, if you can, have the diagram enlarged before cutting the bones apart; many photocopiers, especially those at print shops, can make such enlargements.)

Distribute copies of the rearranged bones, together with scissors, glue, and unlined paper. Tell students that the sheet shows the position of a group of bones recently found in Utah. But paleontologists are stymied: how do the bones fit together? Students can help by cutting out the bones and arranging them to create the dinosaur's skeleton.

Some guidance will be needed here, if only in the form of questions that encourage students to consider *why* they're putting bones in some places and not in others. Don't provide "right" answers though. After all, uncertainty is the norm in paleontology.

When students have decided on their arrangements, have them glue the bones to the unlined paper and display their skeletons to the rest of the class. Encourage students to explain the reasons for their arrangements, and allow

enough time to discuss differing opinions about where a particular bone should go. (Are some explanations more plausible than others? Why?) Now's the time to introduce words like *paleontologist, fossil,* and *inference. Inference* will be the hardest to understand, but the analogy of the detective who reasons from the available evidence should help. Teachers may wish to use identification cards to introduce new vocabulary to hearing-impaired students. And, if signing, they should reference unfamiliar words in sign language as needed.

Finally, encourage students to discuss what the skeleton can tell them about how the dinosaur looked or the way it lived. Students will probably agree that the skeleton provides a good indication about the number of limbs, the presence or absence of teeth, the size of jaws, and the size of the creature itself. What about diet? *Allosaurus'* saberlike, serrated teeth were perfect for tearing flesh and suggest that the creature was carnivorous. (How do its teeth compare with those of *Brontosaurus?*) But what would the skeleton tell us about internal organs, skin, and behavior?

A look at skeletons of different kinds of dinosaurs—*Brontosaurus, Ornitholestes,* and *Stegosaurus,* for example—can help wrap up the ideas you've been discussing. (Now, too, students may want to look through some dinosaur resources [see box] to find out more about the terrible lizards.) Urge students to more sophisticated inferences about what an animal ate and how it got its food. For instance, compare *Ornitholestes'* attenuated forelimbs—with their three-clawed digits—with the forelimbs of *Brontosaurus.* Why might *Ornitholestes* need clawed digits? Why might *Brontosaurus* lack these claws? What would *Brontosaurus'* long neck suggest about where this herbivore got its food? And what about *Allosaurus'* long hindlimbs? Would long strides make for considerable speed? How is that appropriate for a carnivore? Our conception of how the animals moved, how they distributed their weight, and even what type of environment they were suited for and lived in can usually be determined from skeletons.

dence, and scientists assume that their models are not permanent but will change as new evidence is uncovered.

Dinosaur study is among the most tentative of sciences. Descriptions do exist, but as no one has ever seen a dinosaur, the descriptions are reconstructions based mostly on skeletal remains and on fossil prints of skin (or in the case of *Archaeopterix,* featherlike coverings). Furthermore, most skeletal remains are scattered about when they're

More About Dinosaurs

While many resources are available on dinosaurs, the following are especially helpful for a discussion of inferring about dinosaurs. Several treat misconceptions about dinosaurs and describe the work a paleontologist does, while presenting paleontology as an active science in which discussion and change are expected. They're absorbing reading as well as useful classroom references.

Cobb, Vicki. *The Monsters Who Died: A Mystery About Dinosaurs.* New York: Coward-McCann, 1983. Explains how scientists use fossils to make inferences about the appearance and behavior of dinosaurs (Grades 4–6).

Cohen, Daniel. *Monster Dinosaur.* Philadelphia: Lippincott, 1983. Presents findings about dinosaurs. Also describes paleontologists' conceptions and misconceptions about dinosaurs, showing scientific study

as carried on by individuals who do make different inferences from their data (Grades 5–8).

Diamond Group. *A Field Guide to the Dinosaurs.* New York: Avon, 1983. An illustrated dinosaur encyclopedia containing information on the evolution, environment, and fossilization of the dinosaur (Grade 6 and up).

Freedman, Russell. *Dinosaurs and Their Young.* New York: Holiday, 1983. Describes the family life of duck-billed dinosaurs. Recent fossil discoveries of eggs and hatchlings are interpreted to suggest that many dinosaurs carefully supervised the development of their offspring (Grades 2–4).

Sattler, Helen Roney. *The Illustrated Dinosaur Dictionary.* New York: Lothrop, 1983. Contains entries for all known dinosaurs and other animals of the Mesozoic Era and lists locations of various dinosaur discoveries.

From *Stegosaurus* to *Gallus Gallus*

By now your students have probably begun to realize that a paleontologist's model making, though fascinating, is no easy task. An exercise with some chicken bones will further impress on them the complexity of the work these scientists do.

Boil the meat off some chicken carcasses you have saved in the freezer from your own or friends' Sunday dinners. Then scrub the bones with soapy water and rinse them clean. After the bones have dried, place complete sets in shoe boxes and distribute the boxes to your students. Don't tell them that these are chicken bones; simply ask them to sort through the bones to try to arrange them. Students will find this difficult—the bones are small and need to be wired together in a standing position before they convey a sense of the animal's structure. Think what paleontologists

do when faced with the skeleton of the 77-metric ton *Brachiosaur!*

As a follow-up, have students actually put the skeleton together. ESS's *How to Make a Chicken Skeleton* and *Teacher's Guide for Bones: Identifying Bones and Assembling Animal Skeletons* will both be helpful.

Making Tracks

Dinosaur tracks are further evidence of what dinosaurs were and how they lived. Tracks that are far apart, for example, may indicate a long-legged animal with a long stride. Heavily indented tracks suggest an animal who weighed a lot. For example, a study of *Brontosaurus'* tracks led scientists to conclude that the creatures were suited for slow rather than rapid movement. (The tracks indicated that as the animal lifted one foot to step forward, the other three feet stayed firmly planted on the ground.) Tracks tell how many toes the

animal had and whether it had claws. The presence of many similar tracks in the same area suggests that a large population of dinosaurs lived there, while the fact that most tracks are found in shale (rock formed from the consolidation of clay, mud, and silt) suggests that the dinosaurs may have lived in a swampy environment.

To help students understand how paleontologists make such inferences, plan a unit on tracking in your class. Have students walk through mud, snow, or sand, and then examine the tracks left behind. Can they match the tracks with the students who made them? What can they infer about a person from his or her tracks? (Weight, length of stride, and gait will perhaps come to mind.) What can't they infer? Does standing in one spot produce a noticeably different track? What sort of track is produced by sliding? Hopping? Running? Could they tell what a dinosaur was doing—breeding, fighting, grazing—by examining its tracks?

Encourage students to look for skid marks on a street, bird tracks in the snow, ridges left by a leaf blowing through the sand, tracks left by paint dripping off a paintbrush. Each environment has a unique combination of tracks, and by studying these tracks students can begin to appreciate what sort of inferences scientists can make from fossilized dinosaur tracks. ESS's *Tracks Picture Book* and *The Teacher's Guide for Tracks* can help you develop other activities.

A Journey Through Time

Suppose some aliens landed on Earth and discovered not our bones but some of our possessions. For example, suppose they discovered only a softball glove, a rake, and a picnic table. What might they infer about us? What model would result? Encourage a variety of answers, however fanciful, before you draw the students' ideas together into a model. For example, the aliens might conclude that the softball glove was an article of clothing, probably worn on an appendage, like the hand, which was delicate and needed lots of padding for protection. The rake would perhaps have been used to comb the thick hair covering our bodies. And since it was such a large comb, the aliens might infer that we were very big creatures. The picnic table, then, would obviously be a seat with a plank for our feet and would be appropriately sturdy for such big creatures. Like the paleontologist studying dinosaur fossils, the alien now has a tentative model of human beings. It's inaccurate, but it's logical, considering the limited information available. Pursue the question of making tentative assumptions further. What sort of information would the aliens need to build an adequate model of life on Earth?

A Dinosaur Portfolio

To conclude your unit on terrible lizards, have each student prepare a portfolio on a dinosaur of his or her choice. Students should gather as much information as they can on their dinosaur, using the books you have in class and perhaps finding others in the library. A description of the creature's size, weight, internal organs, skin, skeleton, eating habits, and other behavior will be needed, as well as a drawing or model of the dinosaur. Students should record where they have made inferences and where they've relied on facts. Finally, they should develop a summary statement that shows how tentative their conclusions are.

Your students won't be able to answer all their questions about dinosaurs, but, after all, scientists are still arguing over whether the creatures were warm- or cold-blooded. And no one is sure why dinosaurs disappeared in a comparatively short period of time after long being the dominant form of life. By the end of the unit, though, students will know a lot more than they did about their favorite dinosaur. They'll also have a good idea about how scientists draw inferences from available information and construct models.

Cynthia Szymanski Sunal is associate professor of curriculum and instruction at West Virginia University, Morgantown. Stamp art by Bry Pollack.

Model of the Earth's Forces

Compression, tension, and shearing are three forces that affect the Earth's crust. These forces may act quickly or very slowly. They may involve thick layers of materials or relatively thin layers and may have different effects on different types of rocks. The effects of the forces on the Earth's crust are difficult for students to visualize. However, play dough can be used to model rock behavior with the students supplying the forces. The models will not look like real rock layers, but they will behave like them. As students apply compression, tension, and shearing to the play dough, they can see how the crust of the Earth reacts to these forces.

The play dough exercises will be more meaningful to the students if you first show them slides that illustrate each of the following: fractures, joint systems, faults, drag folds, anticlines, synclines, monoclines, fold mountains, and fault block mountains.

Making "Rock"
To prepare the play dough, mix the following ingredients in a large pan.

- 500 g flour (4 C)
- 250 g salt (2 C)
- 40 mL cream of tartar (8 tsp)
- 60 mL salad oil (1/4 C)
- 1000 mL water (14 C)

Cook the ingredients, stirring constantly, over a low heat until the play dough pulls away from the sides of the pan. Add food coloring if desired. Then knead the dough until it is smooth. Store the dough in a tightly sealed plastic bag so that it doesn't dry out. When you are ready to experiment, fill one petri dish with dough for every two students.

Earth-Shaking Experiments
Before you begin the experiments, give your students a laboratory sheet so that they can describe their procedures and record the results of the experiments. On this sheet, the students should describe the force applied to the play dough and then draw and define the resulting structures.

Compression: For their experiments with compression, have the students form the dough into a rectangular shape. They should apply different degrees of force by pressing their hands together slowly and then rapidly from different angles. They will see that pressing rapidly with much force creates "mountain ranges" in the dough, while slow, subtle pressure causes synclines and anticlines. Also have them experiment with different thicknesses of dough.

Tension: Have your students again form a rectangular shape with their dough and then have them slowly pull on both ends of the form. They should find that the results of this tension are fractures and gaping holes which will vary in shape, size, and number according to how much tension they apply. Have them pull only on one side of the shape to see if there is any difference in structures formed.

Shearing: To imitate the force of shearing, have the students move one hand up and one hand down as they hold on to each end of the rectangular play dough. With this shearing motion drag folds and/or faults should form. Suggest that they experiment more by shearing slowly and then more quickly.

A Solid Finish

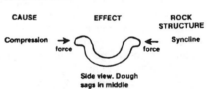

At the end of the experiments show the slides again and ask students to guess what force or forces that may have caused each of the structures depicted. The "correct" answers are not as important as your students relating *their* own results to actual features. Ask your students to explain how the following statement relates to the activities they performed with the play dough: "Nature shows us the results of its experiments; it is up to us to figure out what the experiments were." Your students should have a better understanding of the formations on the Earth's crust. In addition, they will see how much scientific inquiry involves the search for a cause to a particular effect.

Virginia Malone
Earl Rudder Middle School
San Antonio, TX

M & M HALF-LIFE

Most geological processes occur at an irregular and unpredictable pace. In processes such as erosion, deposition, land uplift and volcanic eruption, periods of activity occur in spurts that are separated by long periods of inactivity.

Although geological processes often reveal *relative* time, they do not indicate *absolute* time. For example, we can look at a rock formation and determine which layer formed earlier and which formed later, but we cannot tell exactly how many years ago a particular layer formed.

Testing radioactive minerals in rocks best determines absolute time. Radioactive decay goes on like clockwork, at an even and continuous pace. The nuclei of radioactive atoms break down, releasing particles and radiation. Finally, the radioactive element changes to a stable new element. The radioactive element is called the parent, and the stable new element is called the daughter.

The rate of radioactive decay is measured by half-life—the time it takes for half of the atoms of a parent element to change into atoms of the daughter element. Consider the element radium-226, which has a half-life of 1,622 years. What happens to 10 grams of radium after 1,622 years? Five grams of radium remain, and five grams will have changed into lead.

In this fun lab, you will experiment with a half-life model in which M & M™ candies represent radioactive atoms. The imprinted "M" on each candy represents whether the atom has become stable or not. Students place the candies "M"-side down in a box, shake them, and then count the number of "changed" atoms. The graphs that students produce also make half-life easier to understand.

DARNELL GIRON
Langham Creek High School
Houston, Texas

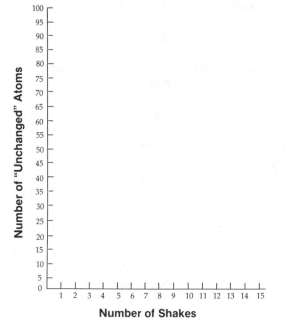

Figure 1. Graph

Trials	Number of "unchanged" atoms
1	
2	
3	
4	
5	
6	
7	
8	
9	
10	

Figure 2. Chart

Materials
- 100 M&M's
- Shoe box with lid
- Graph paper

Procedure
1. Place the candies "M"-side down in the shoe box.
2. Close the cover and shake.
3. Open the box and remove all the "changed" candies (those turned "M"-side up).
4. Count and record the number of "unchanged" candies remaining in the box. Record this data on a chart. (See Figure 2.)
5. Repeat steps 2, 3, and 4 until all the candies have turned.
6. Graph the information from the chart. On the graph (Figure 1), draw a curve in red for the data. In this model of half-life decay, each shake is comparable to the passing of time; the number of "unchanged" candies is comparable to the number of unchanged atoms.

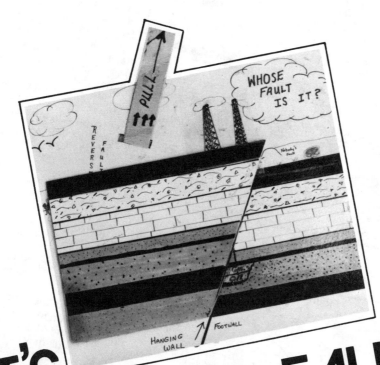

IT'S YOUR FAULT!

The earth imposes the ultimate restraints on human activity and, thus, is an integral part of many social and political concerns. Our planet and its workings are the focus behind such problems as depletion of mineral and fossil fuel reserves, water availability, and our disappearing shorelines. Earth science factors directly impact life choices such as where we might live in relation to potential earth-related disasters.

While these concerns are important, few of us are sufficiently literate about our environment to understand the real constraints it puts on our lives. Earth literacy education, designed to help students recognize their personal roles in relation to the environment, should be made a part of our education system.

Specific Faults

Although activity-oriented earth science programs are necessary for middle school aged students, most teachers do not have the time to develop their own locally oriented science program, and many textbooks do not provide the necessary manipulative experiences. The following activity has been used in classrooms to help students visualize faults and movements that occur during earthquakes. Students make their own models to examine, manipulate, and take home.

An earthquake occurs along a fault or fault zone. A fault is a fracture or an area of fracture along which earth movement occurs. This movement may be only a few centimeters or it could be many kilometers. Faults are divided into two main types—dip slip faults and strike slip faults. A dip slip fault is one in which the slip or relative movement occurs along the plane of the fault (up and down). In a normal dip slip fault (Figure 1), the rock above the fault plane, the hanging wall, moves downward in relation to the rock below the fault plane (the footwall). The hanging wall is the one which would hang over your head if you stood on the fault plane. The footwall is the side of the fault on which you would stand. In a reverse fault (Figure 2), the hanging wall moves upward relative to the footwall.

In the second type of fault, the strike slip fault, the displacement is in the horizontal direction. In a right-lateral strike slip fault, the displacement of the "other side" of the fault is to the observer's right, whichever side of the fault is viewed. The San Andreas fault is an example of a right-lateral strike slip fault. In the left-lateral strike slip fault, the movement is to the left of the observer. Oblique slip faults combine horizontal and vertical movement.

The directions below are for constructing a model of a reverse dip slip fault. The two sides of the fault, the hanging wall and the footwall, are labeled. (See Figures 2 and 3.)

Materials
23 × 28 cm sheet of cardboard
Smaller piece of same cardboard
2 × 50 cm strip of cardboard
Scissors
Felt-tipped pens
Plastic or masking tape
X-acto knives

Earth at Hand

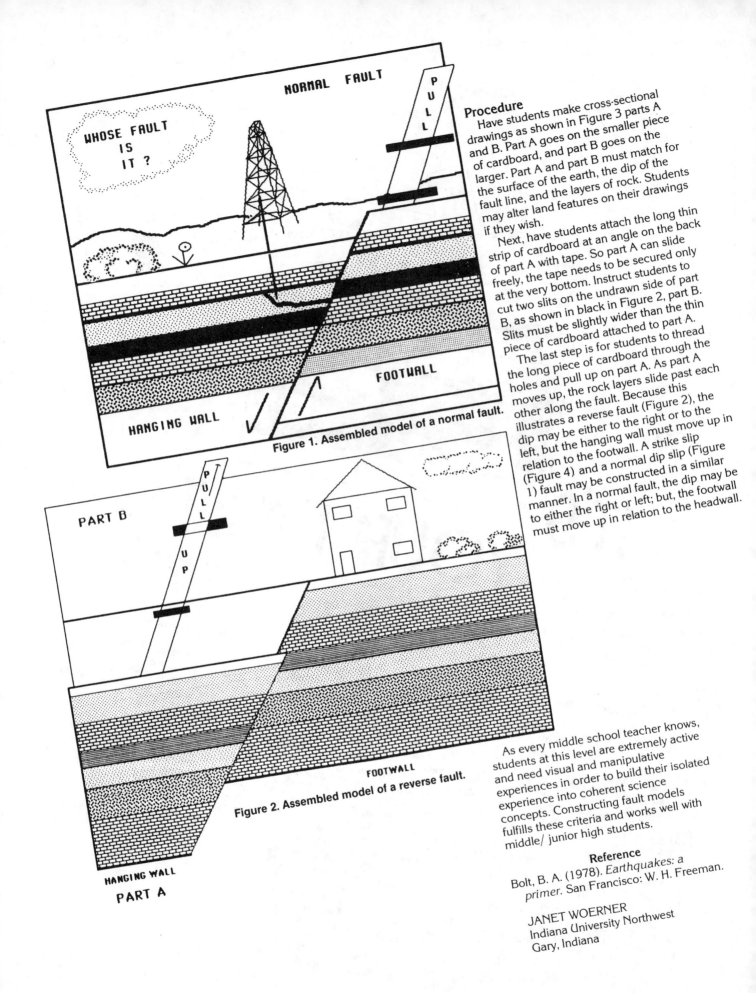

Figure 1. Assembled model of a normal fault.

Figure 2. Assembled model of a reverse fault.

Procedure

Have students make cross-sectional drawings as shown in Figure 3 parts A and B. Part A goes on the smaller piece of cardboard, and part B goes on the larger. Part A and part B must match for the surface of the earth, the dip of the fault line, and the layers of rock. Students may alter land features on their drawings if they wish.

Next, have students attach the long thin strip of cardboard at an angle on the back of part A with tape. So part A can slide freely, the tape needs to be secured only at the very bottom. Instruct students to cut two slits on the undrawn side of part B, as shown in black in Figure 2, part B. Slits must be slightly wider than the thin piece of cardboard attached to part A.

The last step is for students to thread the long piece of cardboard through the holes and pull up on part A. As part A moves up, the rock layers slide past each other along the fault. Because this illustrates a reverse fault (Figure 2), the dip may be either to the right or to the left, but the hanging wall must move up in relation to the footwall. A strike slip (Figure 4) and a normal dip slip (Figure 1) fault may be constructed in a similar manner. In a normal fault, the dip may be to either the right or left; but, the footwall must move up in relation to the headwall.

As every middle school teacher knows, students at this level are extremely active and need visual and manipulative experiences in order to build their isolated experience into coherent science concepts. Constructing fault models fulfills these criteria and works well with middle/ junior high students.

Reference
Bolt, B. A. (1978). *Earthquakes: a primer.* San Francisco: W. H. Freeman.

JANET WOERNER
Indiana University Northwest
Gary, Indiana

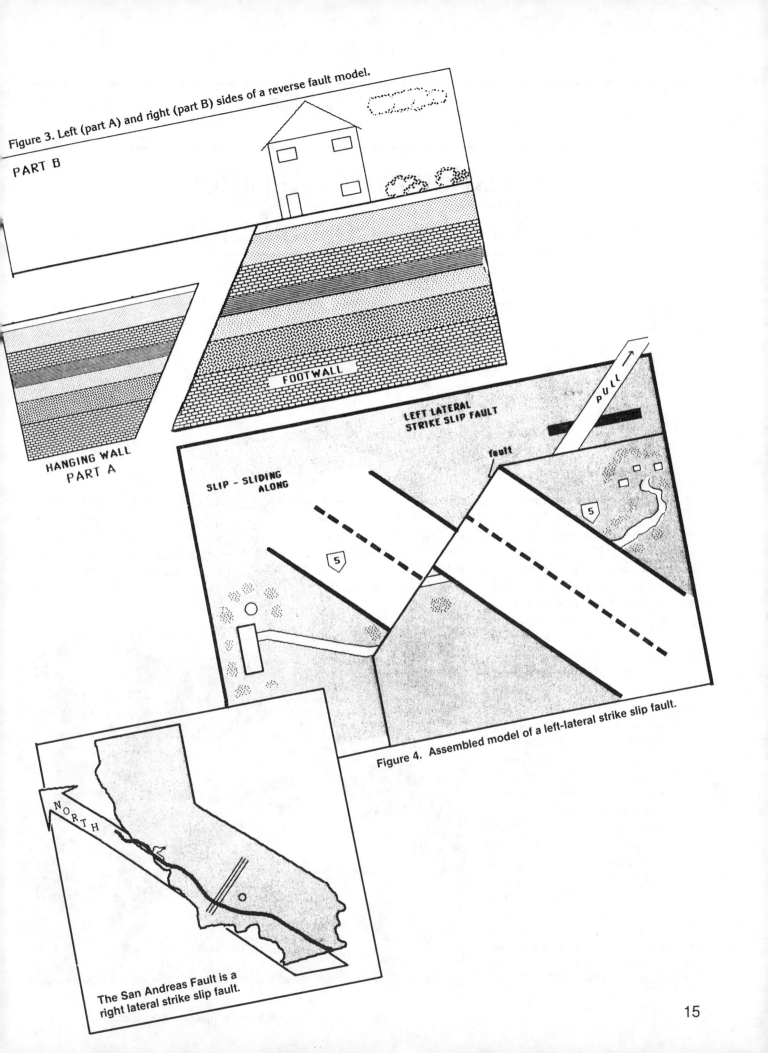

Figure 3. Left (part A) and right (part B) sides of a reverse fault model.

PART B

FOOTWALL

HANGING WALL
PART A

LEFT LATERAL
STRIKE SLIP FAULT

PULL

SLIP – SLIDING
ALONG

fault

5

5

Figure 4. Assembled model of a left-lateral strike slip fault.

NORTH

The San Andreas Fault is a
right lateral strike slip fault.

15

Inexpensive Equipment for a Demonstration of Diastrophism

S. Wilson Tourtellotte

How can you provide effective but economical demonstrations in an introductory earth science course with limited funding? I have faced this problem as the instructor of a biennial survey course for liberal arts nonscience majors. I have found I can save money by using a low cost version of an apparatus known as a Bailey Willis pressure box. This device can help you demonstrate diastrophic movements, the processes that deform the earth's crust to produce features such as folds and fractures. Commercial models cost over two hundred dollars each, which exceeds our budget. My inexpensive version costs less than twenty dollars to build and yields excellent results.

The pressure box basically consists of a large rectangular compartment with a bottom and four sides, one of which is transparent and another movable, and a means for applying pressure to the movable side. The inside

S. Wilson Tourtellotte is an associate professor of physical science at Albertus Magnus College, New Haven, CT 06511.

contains a layered assortment of different-colored, fine-grained, earth-like materials. Pushing the mobile side of the box to compress the materials or pulling it to loosen them will cause the layers inside the chamber to deform,

simulating the events of diastrophism.

After you have built the box, which I will give instructions for later, layer it inside with earth-like materials. Choose fine-grained substances of uniform consistency. You can sift various substances

through a simple kitchen sieve to produce adequate materials. A variety of sources will provide color contrast between layers. This helps make the diastrophism more visible. I have used beach sand, cornmeal, flour, potting soil, sawdust, wood ash, and yard dirt. On occasion, I have ground pieces of red sandstone by hand and sifted them to supply a distinctive red sand.

Sift these materials into the box one at a time in approximately one-half-inch deep layers. You can sift materials directly into the box through the open top section with the movable end panel in place. To demonstrate a number of different diastrophic processes, vary the texture of the layers, the position of the movable panel when the box is being filled, and the direction in which you apply pressure.

Pushing inward slowly and firmly on the movable end will simulate first monoclinal and then anticlinal folding. By applying force in a similar fashion, but with more pressure to the bottom than to the top, you can create reverse faulting, especially if the layers are granular in consistency and the box is filled only to about a quarter of its depth. You can illustrate normal faulting by using cohesive materials such as wood ash or flour and allowing the movable panel to slide outward, and you can produce synclinal folds by using less cohesive, looser layers. An alternative way of making synclines is to produce them during the formation of multiple anticlines.

To create a dome structure, place a deflated rubber balloon attached to a bicycle pump on the bottom of the box. Then add layers of dirt and inflate the balloon. Deflating the balloon will then produce basin formations. Such miniature scale demonstrations illustrate the different types of diastrophism in a clear and tangible way. For the students, seeing is believing. ∎

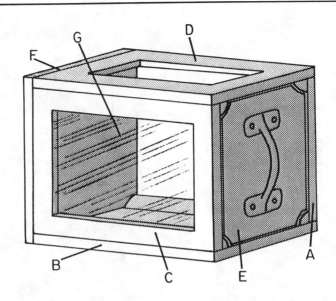

Pressure Box Construction

You will need the following items to build the box: four 12-inch lengths of ½-inch quarter-round molding; one 8 x 12-inch piece of glass or preferably Plexiglas™ for safety reasons; one 2½-ounce package of 5-minute epoxy; one 5½-inch metal pull-handle with screws; a small handsaw and wood rasp; and six pieces of ½-inch thick cabinet grade plywood. Cut (or specify the lumber yard cut) the plywood to the following specifications, where sides correspond to those diagrammed in Figure 1: two 8 x 12-inch pieces, one to serve as side A and one as the bottom B; two 8 x 12-inch pieces with a 6 x 10-inch central section removed, one for the transparent side C and one for the top D; one 6⅞ x 8-inch piece, which will be the movable end E; and one 8 x 9-inch piece to be the fixed end F.

Using epoxy, join sides A and C to the bottom B in such a way that the side panels rest on the bottom piece. Then glue the top section D onto the two side panels A and C. Next, insert the glass G into the box, so it fits up against the opening in side C. Glue the glass in place and reinforce this attachment by gluing lengths of quarter-round molding at the joint between the top and the side and at the joint between the bottom and the side. Glue the glass (or Plexiglas) in place and reinforce this attachment by gluing lengths of the quarter-round molding at the joint between the bottom and the side. Glue the two other lengths of quarter-round molding into the corresponding corners at the back of the box. These four pieces of molding will also serve as tracks to guide the motion of the movable end of the chamber, so be careful to keep their outer curved surfaces free of glue. Fit the movable end E to the box by sawing out quarter round segments from each corner and rasping to adjust the fit to the molding tracks. Screw the handle to the outer face of this movable end in a vertical position. Finally, glue the fixed end F to the outer edge of the box opposite the movable end.

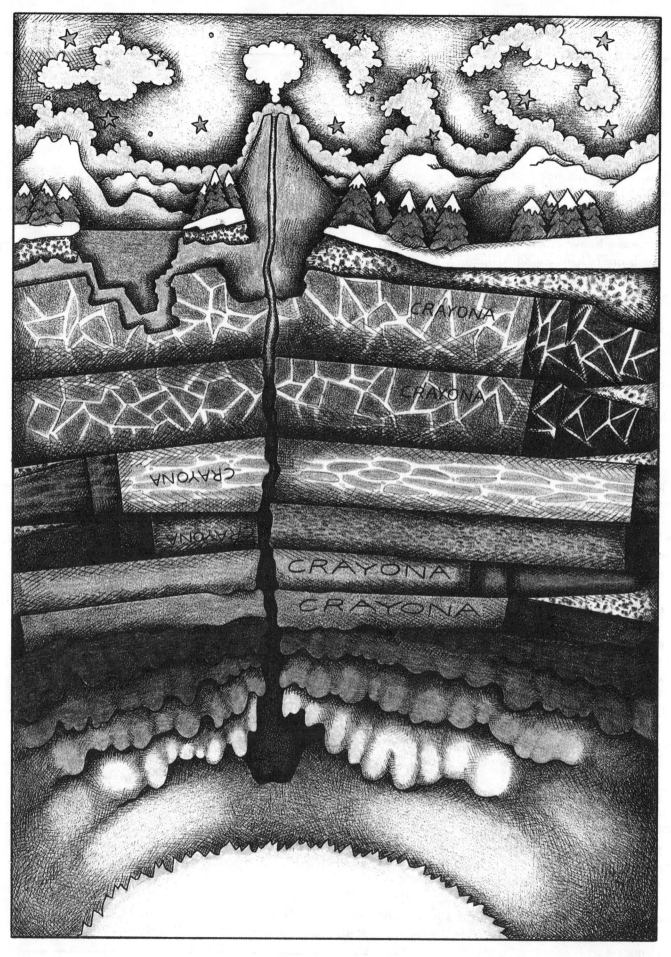

—Art by Robert Schwanfelder

National Science Teachers Association

Color Me Metamorphic

by Donald L. Birdd

Use a box of crayons to illustrate the rock cycle.

Many of our students are not yet at the developmental stage where they can truly understand the abstract concepts presented to them daily. The rock cycle, a major geologic concept taught in every Earth science course, is a typical example.

Your students can "see" the rock examples in the classroom; the difficulty lies in their inability to visualize just how these rock samples were formed. There is no question as to whether or not this concept should be included in a high school Earth science course—but how can we make it easier for students to understand? I have found the following lab extremely effective; it gives students the opportunity to "see" the rock cycle through a series of simulation activities: mechanical weathering and erosional processes, and formation of sedimentary, metamorphic, and igneous rock.

You will need a minimum of two class periods to complete the activity, although I recommend three or four. It naturally becomes an integral part of the rock cycle unit.

Setting the stage

Start by asking your students to describe local rocks and/or rock formations, or ones that they have seen during walks along a beach or river's edge, near or on a mountain, or during drives along highways that were built through road cuts. Be sure to have rock samples distributed around the room.

Now that you have set the stage, ask your students questions: "Have you ever wondered just how these rocks form?" "Are new rocks forming at this moment?" "Why do they break up into small pieces?" "What about the layers . . . and the streaks?" You could ask each student to write down one rock- or rock-cycle-related ques-

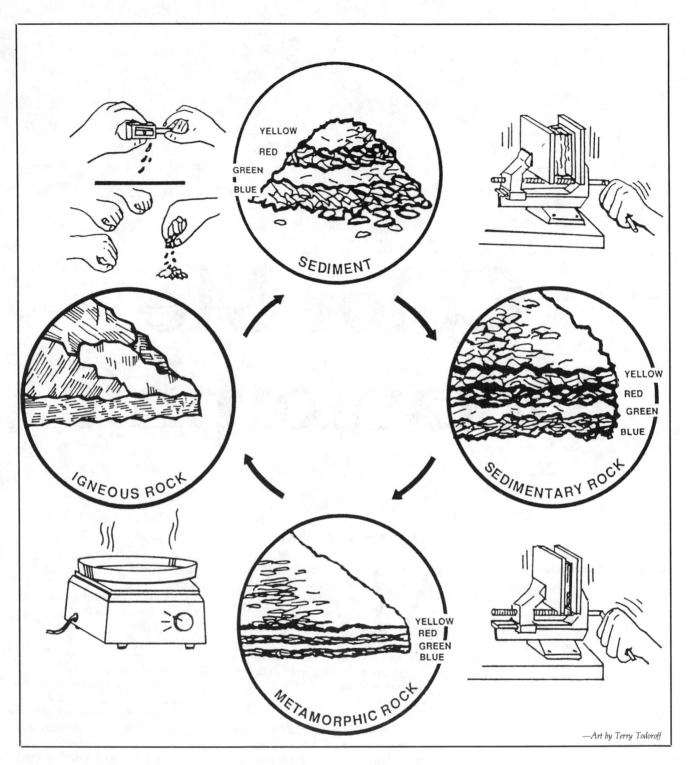

—Art by Terry Todoroff

tion they would like to have answered in class.

Apparatus

Safety goggles
Laboratory aprons
Vise (bring a small one from your workbench or obtain permission to use those available in the school shop)
Board pieces (two per vise, 2.54 cm × 12.5 cm × 20 cm)

Hot plate, flat surface (one per 4–6 students)
Tongs (two pair per group)
Pocket pencil sharpener (one per student)
Petri plates or finger bowls (one per 4–6 students)

Materials

Wax crayons: red, green, blue, yellow (4–6 of the same color crayon per

student)
Aluminum foil, heavy duty (45 cm × 45 cm squares; one per group, and one piece to cover each hot plate)
Aluminum foil disposable pie tray (10 cm diameter or rectangular tray of similar size; two per group)
Aluminum foil to line trays
Wax paper
Envelopes
Newspaper (enough to cover lab tops—have lots of it handy)

National Science Teachers Association

Part A: Weathering
(Lab Period #1)

Cover all desk/lab tops with newspaper. Give each student a sheet of wax paper, a pocket pencil sharpener, and 4-6 crayons of the same color. The crayons represent rock material, and the pencil sharpeners represent mechanical weathering agents. Students should carefully shave each of the crayons with the pencil sharpener, keeping all of the fragments in a small

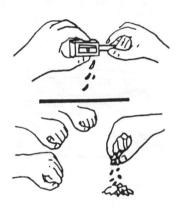

pile. As they are "weathering" their crayons on the wax paper, call their attention to the size and shape of the fragments. "Are they all the same?" "Why or why not?" "What is true about the size of rock fragments in nature?" "Where do they tend to weather more?" "What are some of nature's weathering forces?" "Where do rock fragments tend to collect?" "Why?" "How do they get smaller in size outdoors?" "Why do the similarly sized ones seem to be found together?" (Note: you should discuss chemical weathering processes as the students are finishing this task.)

When the weathering is complete, wrap the fragments in wax paper and place them in an envelope.

Part B: Erosion and sedimentation
(Lab Period # 2)

Once rock fragments have been created, they are usually moved by some force of nature; here, the students act as the erosive force. As they are covering their desks/lab areas with newspaper and retrieving their envelopes of "rock fragments," ask them what this force of movement is called, and what are some of its causes.

Place the four piles in an area accessible to everyone in each group. Each group needs one sheet of foil 45 cm × 45 cm, which must be folded in half (now it is 22.5 cm × 45 cm). Carefully, one student from each group should transfer one color of "rock" fragments to the center of the foil, spreading the fragments into a 1-cm-thick layer covering an area approximately 8 cm × 8 cm. Each color of "rock" fragments should be applied to the foil in a similar manner, creating layers of color.

Focus discussion on deposition and stratification. Some appropriate questions might include: "What forces allow this process to happen?" "What forces sometimes interfere with this process?" "Where could you find layers of rock fragments in loose layers?"

Once students have recorded their observations of what they have collected on the foil, they should fold the foil over the "rock" fragments. CAUTION: Allow for a 1-cm distance between the shavings and the four foil folds.

Part C: Sediments/sedimentary rock simulation (Lab Period #2)

This is a "bottle neck" in the activity. Unless you have three or four vises, this step takes time—everyone must be patient. You might arrange for this part of the lab to be done in the shop area of your school.

Each group places their folded foil package between two boards. The "sandwich" should then be placed in the vise. As (light) pressure is applied with the vise, the "rock" fragments are compressed. This part of the simulation requires that students begin to understand the lithification process—

that cementation accompanies compaction. Spaces between fragments, reduced in size, fill with any of these cementing agents: calcite ($CaCO_3$), silica (SiO_2), or iron oxide (Fe_2O_3). Once the "rock sandwiches" have been mildly compressed, remove them from the vises. Students should carefully open their packages and observe the new product. Call their attention to the central region which is more tight-

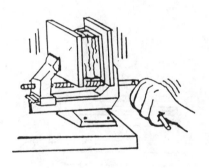

ly compressed; they should lift this portion from the non-compressed fragments and carefully break it into two parts. Have them look at the broken edges and describe the layers. How do they compare to the original layers? What happened to the spaces between the fragments?

Transfer the loose fragments of "rock" into an aluminum tray, leaving the two pieces of "sedimentary rock" in the aluminum package and wrapping it as you did originally.

Part D: Metamorphic rock simulation (Lab period # 3)

Place the foil packages between the boards and in the vises again. Tell the students to add as much pressure to the vises as they can. This part of the activity demonstrates the need for greater pressure. In reality, as the pressure deep within the Earth increases, temperatures increase as well.

A temperature change is probably occurring in this activity but the change cannot be measured—the chemical activity associated with the formation of metamorphic rock is not a part of this activity. It is important for students to understand that metamorphic rock may become contorted in appearance—and actually flow like a plastic material—in response to the pressure that is probably caused by over-riding rock load and plate movement.

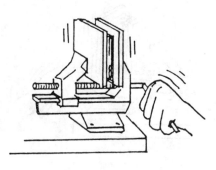

Have your students release the compression on the vices and remove the foil packaging to examine the newly formed "metamorphic rock." They should break the "rocks" open and examine them carefully, noting what happened to the thickness, fragment shape, and surface that was against the foil. Is the latter dull? Shiny? Why?

Place the previously saved "sedimentary rock," some of the "sediment" fragments, and a piece of the "metamorphic rock" in a petri plate or finger bowl for future observations.

Part E: Igneous rock formation (Lab Period #3)

SAFETY NOTE: STUDENT SAFETY IS ESSENTIAL DURING THIS PORTION OF THE SIMULATION ACTIVITY. WHILE STUDENTS SHOULD HAVE BEEN WEARING SAFETY GOGGLES AND A LABORATORY APRON THROUGH-OUT THE ACTIVITY, THIS PORTION REQUIRES THAT THEY BE ESPECIALLY SAFETY CONSCIOUS BECAUSE THEY ARE WORKING WITH THE HOT PLATE AND MELTED WAX.

Igneous rocks form deep within the Earth. They originate in magma chambers embedded in solid rock. These rocks are either extrusive or intrusive in nature—products of volcanic eruptions are extrusive whereas those molten rocks which cool and stay within the Earth are intrusive. We see intrusive igneous rocks only when erosion exposes the previously formed structures.

Each student group should prepare one of the following special activities:

Groups 1 and 2: Line an aluminum tray with foil; fill it with crushed ice.
3: Line an aluminum tray with foil; fill it with ice water.
4: Line an aluminum tray with foil; fill it with warm water.
5: Line an aluminum tray with foil; fill it with hot water.
6: Line an aluminum tray with foil.

If you have more than six groups, decide which activity you want the extra groups to duplicate.

For the "igneous rock" simulation, place all crayon fragments in the aluminum tray EXCEPT those which you have set aside for later comparisons. BE ESPECIALLY CAREFUL HERE! This portion of the activity requires a hot plate as a heat source. AVOID DROPPING WAX FRAGMENTS ON THE HOT PLATE SURFACE. Cover the hot plate surface with a layer of foil before you turn it on. Place the tray on the hot plate and turn the hot plate temperature to medium. Melt the wax, but be careful so the melting process does not occur so rapidly that the molten wax spatters. When most of the "rock" is in the molten state, turn the hot plate off. There is enough heat energy in the molten wax to melt the remaining solid mass. CAUTION: DO NOT LET THE WAX MASS HEAT TO THE SPATTERING POINT!

While the wax is still in the molten state, a student from each group, with tongs, should CAREFULLY:

Group 1: Lift the tray with the molten wax off the hot plate and place it on the crushed ice surface. Place foil over the molten wax surface and cover with more crushed ice.
2: Pour the melted wax over the sur-

face of the crushed ice.
3: Pour the melted wax into the ice water.
4: Pour the melted wax into the warm water.
5: Pour the melted wax into the hot water.
6: Transfer the melted wax tray to the aluminum covered tray. DO NOT POUR OUT THE MOLTEN CONTENTS. LEAVE THEM IN THE ORIGINAL TRAY.

Students should be able to make many observations during this part of the activity. Encourage groups to compare what they did. For instance, comparisons should be made between the "crystal sizes" formed by groups one and six.

Set aside all of your trays until the next day's class; the materials must sit overnight. This will allow the wax mass to cool, and you will be able to carefully remove the solid from the foil. Be sure to look at the lower surface of·the solid wax bar (the "igneous rock"). Discussion will follow naturally.

In group two's activity, the effect the molten wax has on the crushed ice. Notice what the molten mass looks like after the ice has melted. Carefully observe on the second day. Do you see any "crystal shapes?" What do you suppose happens when a volcano erupts in a cold region?

What happened to the molten wax when it was poured in the ice water? Think in terms of an Icelandic volcano. What do you suppose happens to lava on the floor of a cold ocean basin?

The results obtained by groups four and five may bring to mind volcanism in warm Pacific waters. Also, remember, pictures have been taken of molten material that appears to be oozing from the edges of continental plates. How would you describe these "rock" structures?

In conclusion

It is important for everyone to understand that all conditions for rock formation cannot be simulated. In fact, geologists have never "seen" intrusive rocks form. However, they are able to look at all of the available evidence, simulate some conditions in the laboratory, and arrive at similar conclusions.

A final assessment: How did your very concrete-thinking students relate to this activity? And what about even those who you know are abstract thinkers? I suspect that they, too, have a much greater appreciation for the rocks that make up so much of the Earth's surface. ■

Note
An expanded version of this activity will be published as a part of the forthcoming American Geological Institute's *Earth Science Investigations*, a collection of activities for grades 8–12.

Donald Birdd can be reached at Earth Science & Science Education, Buffalo State College, 1300 Elmwood Avenue, Buffalo, NY 14222.

—*Photo by Ken Roberts*

Jello Geology

A discussion of the layers of the Earth often finds students puzzling over two concepts: "How do we know what the inside of the Earth looks like if we have never seen it?" and "Why is the Earth layered?"

While no model can ever be completely accurate, I find that the following activity involving Jello 1-2-3 (a gelatin product that forms colored layers when it sets) helps students better understand the complex subject of earth layering

You will need approximately one packet of Jello 1-2-3 for every fifteen students, a blender, and as many plastic spoons, clear drinking straws, and opaque drinking cups as there are students. Prepare the Jello in class according to the instructions and pour the mixture into the individual cups. Chill the Jello for the amount of time specified (you can also prepare the Jello ahead of time, but then the students will not see the Jello in its original form, before it forms the layers).

The next day give a cup to each student. Ask students to observe the cups and predict how the inside will look. Some students will already know about Jello 1-2-3, and will attempt to "spill the beans." Use this as a point of discussion. Ask them how they know for sure that the Jello will look like, and whether the Jello will always set in the same way. A discussion of this can lead to the conclusion that even scientists are not always sure of the outcome of their predictions. When the students have completed their predictions, show them how scientists investigate the Earth's interior by having them sample their models with the spoons, use the straws to take core samples of Jello, and "thump" the side of the cup, which, by vibrating in different ways, will indicate the presence of at least one substance that is of a different density than the top layer.

Have students note the differences between their predictions and their data. Then ask students to compare the Jello's differential layering to that of molten earth, explaining why each separates as it cools. Help students discover that the Jello forms layers in the same way that the Earth formed layers in its molten state—the heavier materials were pulled to the bottom by gravity. After students have worked up their appetites as "Jello geologists," conclude the lesson and allow them to go on to dessert.

Kenneth E. Raisanen
Ontonagon Jr/Sr High School
Ontonagon, MI

Economics of Earth Science

Science teachers frequently look for ways to incorporate other disciplines into their subject area. You can integrate Earth science with the study of economics so that students become aware of the interdependence of nations and learn about the mineral resource base of our society. To do this, you will need a subscription to a business periodical such as *The Wall Street Journal* or a similar newspaper that publishes daily prices for mineral commodities and future prices for those same minerals. Have students record the prices for selected resources such as gold, tin, Arabian light oil, fuel oil, or coal. Assign a commodity to each student for daily monitoring. Select some students to record the daily comparison in prices for closely associated commodities such as gold vs. silver, Arabian light oil vs. Texas crude, Kruggerand vs. maple leaf vs. gold bullion, or any number of potential groupings.

After sufficient data has been collected, students may graph the information and then attempt to relate changes in prices to world and national events. One of the correlations my students noticed was the drop in the price of Kruggerands during rioting in South Africa last winter. Platinum prices soared when a strike broke out in one of the world's largest platinum mines. They also kept track of the seasonal fluctuation in the prices of fuel oil and gasoline.

In addition to collecting data about prices, students became experts in their particular commodities by researching the sources of their minerals, how they are refined, if there are any substitutes for the minerals, and what they are used for in our industrial world. Students also collected pictures and, in some cases, the actual commodities to include in final reports.

During the course of the study, which took only a few minutes at the beginning of each class period and one or two class periods in the library, my students and I discussed insider trading, supply and demand, market gluts, over-production, future trading, cartels and free markets.

Ken Uslabar
Shattuck Jr High School
Neenah, WI

Eyes on the Earth—*The Frontiers of Remote Sensing*

Pulsing swirls on Jupiter, intricate multicolored rings around Saturn, raging storms on Neptune—scientists and the public marvel at the beauty and mystery of our solar system as revealed in the digital images beamed back to Earth by NASA's unmanned satellites. Likewise, satellites orbiting Earth provide a unique perspective on places and events closer to home: forest fires in the Rockies, hurricanes in the Caribbean, the ripening of winter wheat crops in the Great Plains, desertification in Africa. These satellites carry remote sensing instruments. Unlike a camera, which records an image onto photographic film, electronic scanning systems used on satellites record images in digital or computerized form. This digital form facilitates computer processing, as well as radio transmission back to Earth.

Unlike the human eye, remote sensing instruments are not limited to the visible energy we call "light." These instruments can sense and record the invisible: ultraviolet energy, infrared energy, and microwaves.

Remote sensing instruments provide more than "pretty pictures." Satellite remote sensing provides the only feasible means to monitor large regions of the Earth at repeated intervals. From the patterns displayed in these images, scientists hope to learn more about the underlying processes controlling the Earth's land, air, and water.

Visible and near-infrared wavelengths are useful in mapping the types and amount of the Earth's vegetation. Scientists have used satellite images to monitor deforestation in the tropics. Thermal infrared wavelengths are used to track the movement of the Gulf Stream as it meanders across the North Atlantic. Instruments sensing microwaves (radar) are used to map ice thickness and monitor yearly changes in our polar ice caps. The list of applications is continually expanding.

Remote sensing has stimulated a holistic view of our planet, underscoring the linkage between land, air, oceans, and inhabitants. Environmental monitoring through satellite remote sensing will play a key role in developing a strategy to insure the health of our planet. Monitoring can only tell us changes are taking place; people must also have the will to halt or reverse these changes.

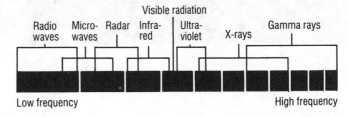

ACTIVITY

This activity engages students in interpreting the information provided by an actual remote sensing image of the United States at night. Students can work individually on the activity, or it can be used as homework. Allow 30–35 minutes for this activity.

MATERIALS

■ Copy of the "USA at Night" image
■ Blank maps of the United States
■ United States atlas
■ Tracing paper

PREPARATION

Imagine you are aboard a spaceship from the planet Xenon on a voyage of intergalactic exploration. The spaceship is equipped with remote sensing instruments that continuously scan your surroundings.

Take a look at the remote sensing image that the spaceship's computer has just generated. Where are you? What are you looking at? Clusters of stars, distant galaxies? No, you're on the dark side of planet Earth, looking at the North American continent at night. Electric lights illuminate the night sky, revealing the locations of major population centers. You already know a little about this planet Earth (you've intercepted television broadcasts of old reruns of "Mr. Ed: The Talking Horse" and "Gilligan's Island"). What else can you learn about these Earthlings, especially those called Americans, from this image of the United States at night?

1. Using tracing paper, draw an outline of North America. Is it easy to distinguish land from water? In some places and not others? Can you draw in the borders of the United States? Label the countries to the north and south.

2. Why do you think the

eastern portion of this continent is brighter than the western coast?

3. Many U.S. cities started and grew up along natural and manmade transportation corridors: rivers, canals, railways. On your map, draw in the Mississippi River, Hudson River/Erie Canal, and Missouri River.

4. You need to replenish your water supply. Where are there large freshwater lakes that might serve this purpose? Draw and label.

5. During the earlier years of American history, people relied on falling water to power their mills and factories. Cities sprang up along the "Fall Line"

where rivers from the hilly Piedmont "fell" down to the flat Coastal Plain. This "Fall Line" is especially evident in the Southeast United States. Draw it in.

6. In the Midwestern portion of the continent, the number of cities thins out as you move west. What natural feature might explain this phenomenon? Label on map.

7. In the Western United States, many cities are strung out in a north-south direction. What natural feature might explain this phenomenon? Draw and label on your map.

8. From what you know of North American geography, label the major cities of the United States and Canada.

Why might Mexican cities not show up so brightly?

9. Once you've landed your spaceship, you make your way to a nearby gas station and request a road map of the United States. How does this map differ from the map that you've drawn using the remote sensor?

EXTENSION
Have students create what they think a remote sensing image would look like if focused on another part of the world at night. Use a world atlas as a reference.

This Lab Is Garbage!

A Take-home Activity

One of the serious problems facing our nation is dealing with our solid wastes. It is important to understand the sources and types of waste to help solve the problem. The purpose of this activity is to analyze the solid wastes generated by your household to think about ways to help reduce that waste.

You will analyze your garbage for a one-week period. During this time you will aim to find out how much is generated and what types of wastes are generated. You can measure the amount in pounds by using a bathroom scale (place the garbage bags directly on it). To determine how much of each type of waste is generated you should separate the garbage throughout the week. Find separate containers to keep the following: paper products, plastic, glass, aluminum, and metals.

At the end of the week determine the number of pounds of each type of material generated as well as the total pounds generated for the week. The following questions should be included in the lab report:

- How much solid waste is produced by your family in one week? How much is this per-person per-week?
- How much solid waste does your family produce in one year? If we assume the solid waste generated per-person in your family is typical of the amount produced by other families in this area, how much waste is produced by the people in this city/town?
- How is your family's solid waste disposed of? Explain where it goes and what happens to it.
- Determine what percentage, by weight, of your family's solid waste is each of the following types: glass paper, plastic, aluminum, metal, and "other." What might be included in the "other" category?
- How much does it cost your family per-pound to dispose of these wastes?
- Do you recycle any materials? If so explain what and why. If not, explain why not.
- How could the amount of solid waste your family currently disposes of be reduced? Explain at least three different methods.

Chuck Wheeler Handlon
Mankato, MN

Can Plants Break Rocks?

I ask my students the following questions: Have you ever seen plants growing out of the sidewalk or the side of a building? Do those plants have the strength to break up the sidewalk or the rocks that they are growing in? Then I tell them that the activity we are going to do will give them some idea of the ways that plants cause rocks to crack up.

You need dried lima beans, red beans, or corn seeds; two small flowerpots and soil; plaster of paris; and a round piece of window glass the diameter of the top of the pots.

Have your students soak 6 to 10 beans or seeds in water and let them stand overnight. Then plant the seeds just under the soil surface in both pots. Water them. Then have students cover the soil in one pot with a 1-cm-thick layer of plaster of paris, and cover the other pot with the glass. Students then observe and examine the two pots daily. I ask them why we soaked the seeds before planting, what happened to the plaster of paris after 5 days, and what they observed about the glass after 5 days.

We then see if we can find places where plants have broken through asphalt, brick, or cement around the school. We compare the strength of sprouting seeds with the roots of grown trees.

Bill White
Thomas A. Edison School
Indianapolis, IN

Sweet Crystals

When studying intrusive igneous rocks and crystal formation, demonstrate to students how crystals actually grow. Slow growing crystals such as alum can take so long to grow that your solution is likely to spill, dry out, or become contaminated. A safe, easy, and completely portable way to grow crystals involves crystallizing sugar from a solution of sugar, gelatin, and water.

Prepare your solution as follows, doubling the recipe to provide more crystal growth samples. In a saucepan, dissolve one packet of unflavored gelatin in one cup of water. When the gelatin is dissolved, heat the solution to boiling and add two to three cups of sugar. Add one cup at a time, stirring until the sugar is dissolved. After adding the second cup, slowly add more sugar until the solution becomes saturated. Bring the mixture to a boil and continue boiling for five minutes. Next, remove the pan from the heat and allow any undissolved sugar to settle to the bottom of the pan. Pour the clear, top liquid into two (four, if you've doubled the recipe) clean glass pint jars and seal them. Then wrap each jar in a towel to keep it warm.

The next day, remove the towels and observe the jars over several days. You will notice crystals developing in the gelatin matrix. As time goes on, the crystals will grow in size and begin to interact, interfering with one another's growth patterns, mimicking the way intrusive crystals form. The growth may continue for several weeks or months. Have your students do the following:

- record the date sugar crystals can first be seen
- draw the crystals to show their growth rate
- describe how the crystals grow
- build their own model sugar crystal
- name the crystal system to which the sugar crystal belongs

The amount of sugar in the original solution determines how long it will take before crystals begin to appear. A well saturated solution will begin crystallizing more quickly than a solution made with less sugar. When you are through with the sugar crystals, discard them or save them to remelt for another crystal-growth demonstration.

Annette Plunkette and Ken Uslabar

Making and Mining a Mountain

Sarah E. Klein

An earth science activity that I have used to stimulate students' interests and challenge their ingenuity calls for some salt, sawdust, sand, and iron filings. I mix these granular ingredients into a gray mound and leave it on my desk before class starts.

After initial student comments, "What's that?" "A new cereal?" "Gerbil cage litter?" we chat momentarily about materials and I describe what we are going to do.

On the chalkboard, I list the materials involved. As we begin describing and examining the physical and chemical (in general) properties, we chart the data collected, including the methods of extraction that can be applied to each material as a result of its properties. (See chart.)

Then I ask each student to imagine being the person in charge of research

Sarah E. Klein is President of NSTA, 1981-82. She is a Seventh/Eighth Grade Teacher at Roton Middle School, Norwalk, Connecticut. Artwork by Johanna Vogelsang.

Materials in a Mountain		
Ingredients	**Properties**	**Extraction Method**
NaCl	White, granular, dissolves in water	Flush material with water, heat, or dehydrate runoff
Sand	Brown, granular, sinks in water	Place material in water, collect residue
Iron filings	Black, attracted to a magnet, sinks in water, oxidizes in water	Use magnet on material in dry form
Sawdust	Beige, floats in water, burns	Place material in water, collect floating matter, dry. Burn-off of the sawdust is possible if the sample is not required for quantitative measure

and development for a large company. The company has an opportunity to purchase a nearby mountain, which contains extractable resources that would benefit the community and the company. As research experts, we have been put in charge of the project. What should we do?

Hypotheses abound. We talk about mining techniques, and the environmental and social changes that would come about. Sometimes we role play so that students are able to test their ideas. We create a business/community council, debate, and bring in expert testimony from "scientists." Special effort is

made to relate the activities to local government regulations and operations. Who decides how the mining will be done? Who will clean up? Who will profit?

The discussion and debate excites interest and urges the students on to test mining theories and techniques. All they need are mountains.

Building the Mountains

To make mountains, we use cone-shaped paper cups, plaster of paris, water, and some of the gray mound. The amount of plaster of paris is optional. (It will wash away in the mining process.) Mix enough water into the materials to form a thick paste. Too much water will permit the sand to sink and settle at the bottom of the cup, giving the mountain a sandy top.

I suggest to students that they place some "buried treasure" in their mountain such as a ring, marble, rock (boulder), or other retrievable item. The practice discourages the students from hitting the mountain too hard as they mine. Pour the paste into the cups, label them, and balance them in empty baby food jars or beakers to harden and dry.

Someone usually comments on "sweating" of plaster walls as they dry. Temperature readings of the drying mixture inspire discussions about chemical change and heat energy. Record and compare temperature readings. Can students explain why there are temperature differences?

Once the mixture hardens, students invert the cups and examine their mountains. As the mining process begins, we talk about cleavage, hardness, crystal shapes, and sedimentation.

All students want to try their mining ideas and methods—chipping, grinding, reclaiming their treasures. They care-fully hammer the mountain into chunks. Caution the students to use the flat side of the hammer. To pulverize their "mining loads" students use a mortar and pestle and apply whatever technique they can to sort the powder. Be sure to have students cover the fragments with a dry paper towel during the pounding stage to protect their eyes.

The classroom is a beehive of industry. Piles of debris collect and dust and chatter fill the air. Eventually costs and manufacturing problems are examined and discussed and a great deal of science-related learning takes place.

If you want to focus on the quantitative aspects of this activity, you may wish to measure out the amounts of each ingredient used and amounts retrieved after the mining process. The limits of this activity are the limits of your imagination. Try it.

Panning For Gold

During a unit on minerals of economic importance, I discuss prospecting with my students. One method that is of interest to them is panning for gold, probably because this activity is considered romantic, of historical importance, and something students can do themselves. After discussing origins of lodes, or ore deposits, and subsequent placers, concentrations of ore developed by alluvial process, we do a lab using simple materials to simulate panning for gold. By following the procedures listed below, you can perform this activity with your students.

Fill lab trays with about 4 L of clean sand, and elevate one end of each tray with erasers. If possible, use sand from a river. Pure white sand should not be used as color variations are necessary. Add water to the pan until it is 2 cm above the sand. Crush pyrite and galena into pieces no larger than 3 mm in size to simulate natural gold deposits. Sprinkle each lab tray with the simulated gold. Use disposable pie tins as gold pans. The sides of the pie tins are steeper than gold pans, but they work well.

In the prelab demonstration, show students how to pan for gold. You will have to practice this a few times before demonstrating to your class. While holding the pan level in the water, rapidly swirl it in small circles. The load will "float," allowing the more dense materials to sink to the bottom. Next, slowly tip the pan while rotating it to allow the lighter material to float over the edge. During the final stages of tipping and rotating, heavier materials will settle to the bottom. Don't tip the pan too steeply, or the "gold" will be lost. If done properly with river sand, the simulated gold, any real gold, and a black sand containing ilmenite and magnatite will accumulate on the bottom. The latter can be removed with a magnet.

This portion of the activity often leads to a discussion of magnetic sorting, as well as density sorting, both of which are used by miners to separate minerals during processing. During the lab process, students are challenged to produce as clean a pan as possible by removing all but the dark sand. Students are allowed to keep all the "gold" they find, be it galena, fool's gold, or, if river sand is used, possibly real gold.

Frank Ireton
Base Jr. High School
Mountain Home, ID

OIL: PRODUCTION, CONSUMPTION, AND RESERVES

The role of natural resources in the Gulf War

by *Marsha Barber*

Since the August 2, 1990, invasion of Kuwait by Iraqi forces, the world's political and economic attention has been focused on the Middle East. A quick glance at the data in Figure 1 will explain why this region is so important to industrial nations around the world. Iraq has the second largest known oil reserves in the world (100 000 000 000 barrels), while Kuwait holds the fifth place position (91 920 000 000 barrels). By invading Kuwait, Iraq gained control of its neighbor's enormous oil reserves and, consequently, hoped to extend its influence over the international petroleum market.

An international oil embargo stopped oil exports from both Iraq and Kuwait in late summer. The remaining OPEC (Organization of Petroleum Exporting Countries) members, particularly Saudi Arabia and Venezuela, have managed to increase their production by one million barrels per day, compensating for the loss of oil from Iraq and Kuwait.

The price of a barrel of oil in U.S. markets prior to the invasion was below $18. Following that event, the price rose, for a brief period, to a level above $40 (you can find the current price of crude oil in the business section of your local newspaper). By late fall, Saudi Arabia and Venezuela had increased production sufficiently to supply worldwide markets, while at the same time, market demand had fallen. Consequently, the price of crude oil has dropped, but the market undoubtedly will continue to fluctuate as the war continues.

Every American has been affected by the Iraq invasion. During the last few months, we have all paid more to operate our automobiles. In addition, many of us are emotionally involved as we have friends and family involved in Operation Desert Storm. Oil plays a major role in determining the direction of international political and economic decisions.

THE SCIENCE BEHIND THE HEADLINES

In light of the current political situation in the Middle East and the fluctuating prices at the gas pump, it is apparent that students would benefit from a better understanding of the international oil market. During the summer of 1989, I and three other high school science teachers had a unique opportunity to work at Amoco Production Company in Denver, Colorado. We worked alongside Amoco geologists, geophysicists, and technicians to design a new set of curriculum materials for junior and senior high students. By combining the expertise of the geo-

science professionals with the classroom experience of the science teachers, a valuable set of materials was produced.

Under the leadership of the Colorado School of Mines, Amoco Production Company, and the Colorado Alliance for Science, the Denver Earth Science Project was established. The major goal of the Project is to develop curriculum modules (from two to six weeks in length) that address critical Earth science issues. To make the data and materials in the module as accurate and up-to-date as possible, the modules

---

In addition, many of us are emotionally involved as we have friends and family involved in Operation Desert Storm.

---

are designed and written in partnership with industry, business, and governmental agency personnel.

After one year of operation, the Project has successfully piloted and edited two modules. The "Oil and Gas Exploration" module was developed at Amoco in partnership with the Mineral Information Institute, Jefferson County School District R-l, and Adams County Five-Star School District. The second module, "Ground Water Contamination," was developed with the assistance of the U.S. Geological Survey and Denver Public Schools, with major financial support from the

Denver Foundation and Colorado School of Mines.

The next phase of the Project will involve delivery of the first two modules to teachers and continued development of additional modules. Some of the topics being discussed for future modules include Hazardous Waste Management, Earthquakes and Geologic Hazards, Air Pollution, Paleontology and Dinosaurs, and Geography and Mapping. One of the unique characteristics of the program is that the modules integrate Earth science, geography, math, and social studies in hands-on activities that stress "real world" issues.

GULF WAR AND GASOLINE

Our "Oil and Gas Exploration" module includes an activity entitled, "Pump It! Burn It! Ship It!," that would be a valuable addition to almost any science or social studies class, as it deals with the geography of "who has the oil." If students know where the oil reserves are, and who the large consumers are, they will better understand why the price at the pump fluctuates so dramatically, and why there is political unrest around the world.

For this activity, students will need a world map (the larger the better) that has each country's borders outlined, a centimeter ruler, colored pencils (if available), a world atlas, and the data found in Figure 1. Upon completion of the activity, students should be able to describe the worldwide geographic distribution of petroleum, identify and locate the countries that make up the Middle East, and name countries that import oil to meet their domestic needs.

TRACKING THE FLOW OF OIL

This activity consists of three steps:

1. Petroleum geologists use the term "proved reserves" to indicate how many barrels of oil remain in a known oil

field (1 barrel = 42 gallons). Once an oil field is discovered, a number of wells are sunk so the oil can be extracted efficiently. Depending on the size of the field and the type of reservoir, it may take anywhere from a few years to a century to extract all the available oil. At any given time, an oil company can estimate how many barrels of oil a country will produce from the fields it has already found. The number of reserves does not include new oil that

FIGURE 1. Oil at a glance. Estimated proved reserves and oil production statistics taken from the *1989 International Petroleum Encyclopedia* p. 250, 251; worldwide crude oil consumption data was obtained from the *1988 Statistics Source Book*, p. 243.

Country	Estimated Proved Reserves (1/1/89) (bbl)	Oil Production/Day (Estimated 1988) (bbl)	Consumption/ Day (1987) (bbl)
ASIA-PACIFIC			
Australia	*	552 300	625 000
India	6 354 000 000	631 800	*
Indonesia	8 250 000 000	1 137 500	*
Japan	*	*	4 510 000
WESTERN EUROPE			
France	*	*	1 835 000
Italy	*	*	1 845 000
Netherlands	*	*	695 000
Norway	10 435 000 000	1 069 100	*
Spain	*	*	935 000
United Kingdom	5 175 000 000	2 376 300	1 610 000
West Germany	*	*	2 430 000
MIDDLE EAST			
Abu Dhabi	92 205 000 000	1 012 600	*
Iran	92 850 000 000	2 207 500	*
Iraq	100 000 000 000	2 679 200	*
Kuwait	91 920 000 000	1 254 200	*
Neutral Zone	5 210 000 000	*	*
Saudia Arabia	169 970 000 000	4 708 300	*
AFRICA			
Algeria	8 400 000 000	666 800	*
Egypt	4 300 000 000	851 300	*
Libya	22 000 000 000	1 012 500	*
Nigeria	16 000 000 000	1 358 300	*
WESTERN HEMISPHERE			
Canada	6 786 000 000	1 604 600	1 500 000
Mexico	54 110 000 000	2 527 300	*
Venezuela	58 084 000 000	1 658 000	*
United States	26 500 000 000	8 165 900	15 955 000
COMMUNIST AREA			
China	23 550 000 000	2 690 000	2 085 000
U.S.S.R.	58 500 000 000	12 476 800	9 090 000

*Country was not a major contributor in this category.

PHOTO BY CLARE LUMLEY

may be found in the future, or oil that has already been produced in the past. It includes just what can be produced now and in the future. Think of "proved reserves" as the current balance of a bank account.

To represent the quantity of proved reserves, use a pencil to draw a vertical bar from the center of each country listed in Figure 1. These 20 countries have the highest proven oil reserves in the world today. The height of the bar will represent the quantity of proved reserves. Use a vertical scale of 1 cm = 10 billion barrels of oil. For example: the United States has 25 billion barrels (bbls) of oil. The corresponding scaled bar would be 2.5 cm high. You can shade in your bars with a colored pencil or use some other technique to differentiate between the various bars you will be asked to draw (see Figure 2).

2. Using the same procedure, plot the top 20 countries in terms of oil production. Oil production is measured in barrels per day (b/d) rather than in total barrels. Create another bar using a scale of 1 cm = 1 million b/d to plot oil production. Example: the United States produces 8 276 000 b/d. The scaled bar will equal 8.3 cm. (Be sure to note that the scale has changed!)

3. Next, create another bar to show daily oil consumption using the same scale as in step 2 (1 cm = 1 million b/d).

Upon completion of the map, your students will have a very graphic display of the current oil situation worldwide. The questions that follow can launch an entire class discussion revolving around current events. Hopefully, students will go home and discuss "world politics" and "gasoline prices" with other family members.

DISCUSSION QUESTIONS

Q. List the five countries with the greatest quantities of proved oil reserves. What special name is given to this region of the world?

A. Saudi Arabia, Iraq, Iran, The United Arab Emirates, Kuwait, and the Middle East.

Q. What is the impact on the rest of the world when the oil reserves are concentrated in one area?

A. There will be competition for that oil between the nations that need it. This can lead to political unrest. Countries in that area will have a lot of control over the oil supply and what they can charge for it. Due to greater competition for oil, it will probably have to be transported over great distances, thereby increasing the chance for tanker disasters and oil spills.

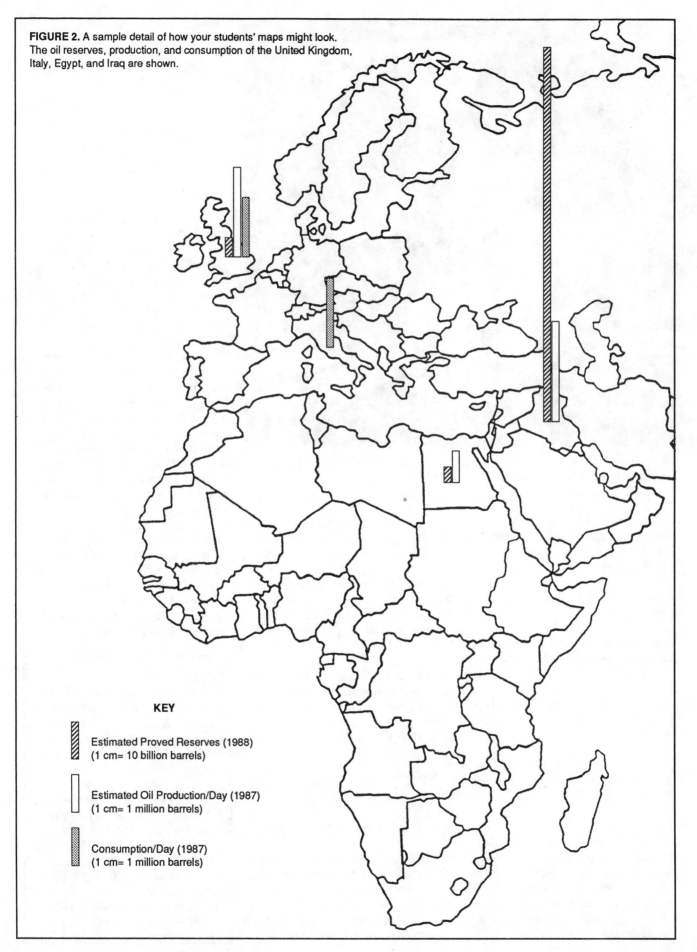

FIGURE 2. A sample detail of how your students' maps might look. The oil reserves, production, and consumption of the United Kingdom, Italy, Egypt, and Iraq are shown.

KEY

Estimated Proved Reserves (1988)
(1 cm= 10 billion barrels)

Estimated Oil Production/Day (1987)
(1 cm= 1 million barrels)

Consumption/Day (1987)
(1 cm= 1 million barrels)

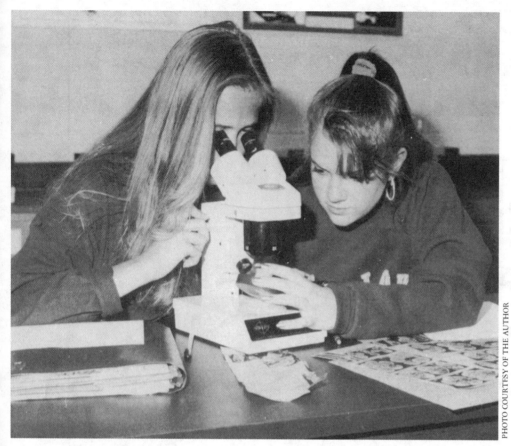

Students studying microfossils as part of the "Oil and Gas Exploration Module" from which this article was adapted.

PHOTO COURTESY OF THE AUTHOR

A. The U.S. consumed approximately 16 000 000 barrels of oil per day in 1987. That amounts to 672 000 000 gallons of oil per day! (Remember there are 42 gallons in a barrel.) As bad as the *Exxon Valdez* oil spill was to the wildlife inhabiting the Alaskan coastline (see sidebar), it only amounted to 1.6% of the daily consumption in the U.S. There are hundreds of similar tankers on the world's oceans trying to meet the international petroleum demand. Most tankers traveling the ocean each day safely load and unload their petroleum at coastal ports. During this discussion, students may also bring up the recent gulf spill created by Saddam Hussein. Initial estimates put this spill at ten to twelve times the size of the *Valdez* spill in Alaska.

Marsha Barber is a physics and Earth science teacher at Chatfield Senior High, Littleton, CO. She is also the Denver Earth Science Coordinator for the Colorado School of Mines, Office of Special Programs and Continuing Education, Golden, CO 80401.

Q. Which countries consume more than one million barrels of oil per day and are able to produce enough to meet that demand?

A. Soviet Union, Canada, China, United Kingdom.

Q. If a country is able to produce more oil than it needs, what does it do with the extra?

A. Export it to countries willing to pay for it; keep it in the ground as a strategic reserve.

Q. List the three countries and one geographic area that use the most oil. Why do these countries require so much oil?

A. United States, Soviet Union, Japan, Western Europe; major industrial nations gererally have a "high standard of living" and need petroleum to power industrial plants, operate transportation systems, and meet the ever-growing energy demands of their citizens.

Q. Which of the countries from the previous question is the most dependent on imported oil?

A. United States or Japan in terms of sheer volume. Since Japan must import all of its oil, it may be considered even more dependent on imports than the U.S.

Q. How is oil transported from one country that has an excess supply to another that has a demand? List three methods of transportation. Which method should we be most concerned about?

A. Truck, train, pipeline, or tanker (ship). Tanker transportation will most likely be of concern to people in light of the *Exxon Valdez* oil spill in 1989.

Q. The *Exxon Valdez* oil spill in March 1989, was one of the largest spills on record. The spill was estimated to be approximately 11 million gallons. How does that compare with the daily oil consumption in the United States?

NOTE

The Denver Earth Science Project was made possible in part by a grant from the National Science Foundation's Private Sector Partnerships to Improve Science and Mathematics Education division.

The authors of the "Oil and Gas Exploration" module are Joe Beydler, Alan Swanson, and Steven Williams. For further information on the Denver Earth Science Project, contact: Marsha Barber, Colorado School of Mines, Office of Special Programs and Continuing Education, Golden, CO 80401; (303) 273–3303.

Politics, Garbage and Fast-Food Franchises

A student survey of trash can lead to increased
understanding of environmental influences.

*Susan Holmes, Ian Lerche, John Pantano, and
Richard Thompson*

Introductory-level environmental earth science courses can be stimulating to students and increase their awareness of the local environment. We have developed several innovative, short (one-or-two-hour) field exercises to accomplish these goals [1,2]. This exercise examines the interplay of three urban environmental factors: politics, garbage, and fast-food franchises. The exercise can easily be extended to other factors, depending on location.

The University of South Carolina at Columbia covers several blocks to the east, northeast, and southeast of the State Capitol. The large student body and large legislative apparatus support a considerable number of fast-food places in the area, ranging from the ubiquitous "golden arches" to the hole-in-the-wall "greasy spoon" establishments. Our initial question was to what extent the urban gutter garbage was dependent on proximity to (1) a fast-food emporium and (2) the State Capitol and USC. Additionally, we wondered whether weekly street cleaning efficiency was a function of distance from the State Capitol, particularly when the

Ian Lerche is a professor of geology and Susan Holmes, John Pantano, and Richard Thompson are teaching assistants, Department of Geology, University of South Carolina, Columbia, SC 29208.

legislature was in session.

The class involved over 130 students, with two formal lectures on the environmental impact of humanity and one related lab exercise each week. The laboratory classes consisted of 20–25 students each, with at least one laboratory class meeting each working day of the week for a two-hour period. (Since neither legislators nor city garbage collectors work on weekends, and students have no classes, it was impractical to extend the exercise over Saturday and Sunday.)

The students' information-gathering technique was as follows: Each lab class was split into smaller groups of two or three and assigned to segments of the three parallel streets bracketing the State Capitol (Assembly, Main, and Sumter), as well as a cross-street (Greene) between Assembly and Sumter. They recorded the garbage they saw in gutters and on the sidewalks along both sides of the street, long with its distance from the Capitol. The manner of recording (five gum wrappers, a pile of leaves, and so on) was left up to the students. The combined notes provided a daily log of garbage distribution for a week. In addition, they recorded the location of the fast-food eateries on their segments of the various streets, weather conditions, and any other related observations.

The breakdown of the "garbage" into categories (bottles, paper, leaves, ciga-

rette butts) varied with the student. Some catalogued everything, while others wrote general remarks such as "a pile of paper" or "lots of trash." This taught students about quantitative versus qualitative input data as they tried contouring data such as "several pieces of junk" to "a pile of cigarette butts" or "seven pieces of paper and one bottle." The process required judgment calls that conjured up the Pygmalion problem. In the end students agreed to accept numerical conversions. (For example, five cigarette butts are equivalent to one popsicle wrapper.)

After an hour or so out on the streets, the students returned to class. For the next hour they pooled their data from the various streets, obtained graphs of the

number density distribution of garbage along each side of the street according to the distance from the State Capitol, and put together a contour map of the gutter-garbage levels in the study area. This exercise teaches them how to collect data, to be aware of local factors influencing the interpretation of data, how to graph and contour data [2], and how to interpret spatial and temporal variations in the graphed and contoured behavior.

Selection of data for plotting on an overlay of the study area led to interesting debates from the students, thereby promoting dialogue and student involvement in the project. (Comments ranged from "I spent more time than you observing so my data are better" to "There is no way we can compare cigarette butts to hot dog wrappers, so let's junk the butts in favor of the bigger wrapper problem.")

The result was a series of six contour maps of levels of garbage, one for each work day of the week and two for Thursday. In addition, the students produced a series of six graphs of the garbage levels on each side of each street plotted according to distance from the State Capitol. At the first formal lecture period in the week following the exercise, the entire class viewed all of the six laboraory classes' contour maps and graphs. After some time to assimilate the similarities and differences, they discussed what the results meant and expressed two major concerns: (1) The daily maps and graphs differed by large factors (probably due to high rain and wind on several days), and their accuracy depended on how diligent and observant each group was; (2) The students debated whether the precision resolution and uniqueness of the contours and graphs would warrant conclusions concerning the political influence factors and/or the street-cleaning factor in the presence of the fast-food factor. (They concluded that the "scruffiness" of their data plots was too overwhelming to show anything except the fast-food factor.) Both the laboratory discussions and larger formal class showed that the students were able to evaluate the resolution, precision, and uniqueness of their project.

The field exercise is valuable for several reasons. It can be carried out quickly and inexpensively in any area, urban or rural, with relevant factors evaluated for their impact. It teaches the students how to collect meaningful data, how to be aware of the "type" of data and the factors impacting on it. The experience teaches students how to select data for further evaluation and provide criteria for quality assessment. They learn how to draw graphs and contour maps, how to interpret them, and how much such instruments can be trusted. Students interpret their results and decide whether they have a sufficient resolution to determine a particular causitive influence in the presence of other, more dominant causitive factors. The exercise gives students an appreciation of a local environmental problem and teaches them to question whether the impact of humanity on the environment can be assessed and communicated to others without loss or distortion of information. In short, they learn to think logically about problems rather than resort to emotional beliefs, which are too often founded on misunderstood statistics and data.

Student reaction to the exercise was mixed. Those who had to venture forth in pouring rainwere very relutant to participate (about 20–30). Others did not want to do any work that involved leaving the classroom (the usual 10 percent). Still others had never learned graphing and contour work and so had enormous difficulties in understanding what they were supposed to do (about 20 percent). Others regarded the exercise as "trivial" (about 20 percent), but most eventually admitted that the exercise was a good introduction to a problem they see and thoughtlessly contribute to every day.

Acknowledgments

We thank John Carpenter for his insightful and critical comments on a previous version of this paper, and Donna Black for her careful typing of the manuscript.

References

1. Carpenter, J.R., *et al.* "The Total Environment: A Confluent Approach to Environmental Education." *Journal of Geologic Education* 24:177–80; November 1976.
2. Carpenter, J.R., J. Dunn, D. Eppler, and D.J. Schultz, "Mapping the Urban Environment with Attitudinal Responses." *Journal of College Science Teaching* 8(2):105–06; 1978.

Dirt-Cheap Soil Studies

One way to incorporate soil science into your classroom is to use the soil surveys distributed by the USDA Soil Conservation Service. These books are free, and available specifically for each state, county, and district. Containing—among other things—aerial photographs of the land, maps, tables of information on the properties and capabilities of each soil type, they are of interest to everyone from farmers to engineers.

The books are basically an information source, and can be useful in a variety of ways; I teach a unit on soil using the book that covers our area.

To begin the unit, I discuss the fact that there are many different types of soil, each having its own specific profile and properties. We talk about methods of determining soil types, and how specific types of soil are put to specific uses.

This is when I introduce my students to the soil survey. I pass out the books—one for each group of two, three, or four students—and ask them to follow along as I locate the type of soil that our school sits on.

First, we look at the map of our county and find our location by following the labeled roads. Once everyone has done so, we move to the Index of Map Sheets that corresponds to the number of the aerial map our school is on.

Next, we turn to the aerial photograph and study it in detail. The soil scientists who put these booklets together use fairly old photographs in order to show the land before too many houses or buildings sprung up to cover it. This can make it difficult to find the area where your school currently exists; however, you can use reference points, such as creeks, canals, and major roads, to locate your exact position.

Now it is easy to determine what type of soil the school stands on. The photograph has lines drawn on it to show the boundaries of each soil type, and each area of soil is numbered. These numbers correspond to a list of soil types in the Soil Legend.

Now that we know the name of the soil, we proceed to learn more about it. Students turn to the "Detailed Soil Map Units" section, which gives detailed descriptions of the soil including depths of each layer, color, natural vegetation, drainage capability, water table depth, and general descriptions of what type of use the soil can be put to.

At this point, I show the students our soil "profile" (an adding machine tape that I have prepared to show what our soil looks like). I have measured and marked off the layers of soil according to the soil description, and colored each layer as accurately as possible. A blue line indicates the average depth of the water table.

After discussing and examining the "profile," we are ready for our field activity. I take students outside to where I have dug a large hole in the ground (and roped it off for safety). We look around at the types of plants in this area before actually examining the soil itself. Hanging the paper "soil profile" over the side of the hole. I compare it to the actual soil. There are usually minor differences in the thickness of the layers, but not too many. We examine and take samples of each soil layer, and compare them to the descriptions in the survey.

Next, we walk to a nearby area where the survey has indicated a different type of soil. I ask the students to compare the vegetation here to the vegetation near the first area, and have them decide if the soil type here might be different. If I have the time, I dig a hole (with student assistance) and look, otherwise we check the soil map to find out if the students are correct in their hypotheses.

We return to the classroom for the rest of the lesson. I ask the students to look in the soil survey for the section of tables that describe the various restrictions on the use of soils. They use the tables to determine if the soil our school was built on is actually suitable for a school; they find information on the suitability of our soil for large buildings, roads, recreational uses such as playgrounds, landscaping, restrictions on sanitary facilities, and even information on how often we can expect to have flooding in our fields.

This leads into discussions about zoning laws, efficient land use, soil conservation, and numerous other topics that the students suddenly find understandable.

To conclude our lesson, I put the students back to work with an activity they can relate to personally: they use the soil survey books to locate their own homes, find the type of soil their home is located on, and do a report on it. They must also construct a model of their soil on adding machine tape. showing the thickness of each layer of their soil and the properly corresponding colors. These models are hung in the room so everyone can see that, although the surface may appear uniform, the soils deep beneath our feet are almost as varied and interesting as the people who study them.

For those students who need enrichment activities, here is an example of some questions they can answer using the soil survey books as the only resource:

Is the soil at your house suitable for growing crops? What type? How many tons of tomatoes could you grow per acre on this soil type? How many horses could you keep if you made your yard into a pasture? What kind of wild plants and trees grow best on this soil?

If these type of activities sound like something your students would enjoy, call your local USDA Soil Conservation Service and talk to the people there. They can help you with everything from getting the books you need to arranging for guest lecturers. You will invest very little of your time, and even less of your money, and your students may stop thinking of "soil" as a dirty word.

Mark E. Kelly,
Suncoast Middle School
North Fort Myers, FL

SOIL LAB:

Groundwork for Earth Science

When budding geologists hear the expression *soil*, they immediately think of *dirt*, something of interest only to farmers and gardeners. The following labs were developed to show students that soil isn't just for growing vegetables. Ordinary soil is a fascinating mixture of very small rocks, mineral particles, living organisms, air, and water. While performing the lab activities, students will become familiar with the different types of soil and their distinctive properties, and also discover how important soil is to everyday life.

In the first lab, students learn the basic types of soil and their characteristics. There are many ways to classify soils, but one of the most basic is by texture. Soil particles can be fine, medium, or coarse. Soils are classified as either sand, silt, or clay, depending on the size of their particles. Sand particles are large and coarse. Silt has medium-coarse particles, and clay is composed of particles so fine they can only be seen with a microscope.

Soil characteristics differ with the size of the particles. Because it is quite porous, sandy soil holds very little water. Clay soil, which feels very smooth and greasy to the touch, binds together well and allows little water to pass through. The drainage properties of silt are intermediate between those of sand and clay.

Loam, the best soil for agriculture, is also sold as potting soil. Loam is a mixture of sand, silt, clay, and organic material which holds water well, allowing a moderate amount of drainage. Loam soils are classified according to the three textures and named for the dominant one, so that we speak of sand loam, clay loam, and silt loam.

Laboratory I: How Does It Feel?

Objective:
Students will experience the different textures of soil and describe them in familiar terms, such as *smooth, rough,* and *soft.*

Materials for each group:
- 15 mL (1 tbsp) coarse sand
- 15 mL silt
- 15 mL clay soil
- 15 mL loam (Use topsoil with a dominant sand mixture for easy identification.)
- paper towels
- magnifying glass
- paper cup of water
- Optional: small straight-sided glass jars (Junior-size baby food jars work well.)

Procedure:
1. Distribute a few paper towels to each group of students and ask them to spread the towels over their work area.
2. Give each group a spoonful of each of the first three soils.
3. Instruct students to rub each sample between their fingers and record their observations on Lab Sheet I.
4. Next, students should look at their samples under a magnifying glass and record observations about the color and shape of the grains and their general appearance.
5. Ask students to moisten the samples and see if the grains crumble, stick together slightly, or mold into a shape. Again, they record data.
6. Give each group a spoonful of loam. Ask them to determine if the loam has a sandy, silty, or clay texture, and have them record their observations.
7. Invite groups to start the optional activity before handing in the lab sheets. The soil samples will form layers in the jar, with coarse sand at the bottom, then fine sand, silt, and clay.

National Science Teachers Association

Laboratory II: How Fast Does Water Pass Through?

Objective:
Observing the drainage capacities of different soils will allow students to see the relationship between soil type and flooding.

Materials for each group:
- 3 styrofoam cups, 180-270 mL
- pencil for punching holes
- waterproof marker
- cotton to cover bottoms of cups
- 300 mL water
- 160-180 g of sand
- 160-180 g of clay
- 160-180 g of loam
- 3 clear plastic cups, marked in mL, large enough to hold the styrofoam cups
- clock with minute hand (All groups can use the large classroom clock.)

Procedure:
1. Provide the following instructions:
- Punch five holes in the bottom of each styrofoam cup, and make a mark 5 cm from its base.
- Put a layer of cotton in the bottom of each styrofoam cup, then add one of the soils to the 5 cm mark. Pack soil tightly, making sure there are no spaces around the sides of the cups.
- Label each cup with the type of soil it contains and place it inside one of the transparent cups. Add 100 mL of water to each cup from the top.
- Determine the length of time it takes for the first drop of water to pass through each soil sample and hit the bottom cup. Measure the amount of water in each cup at 2-minute and 5-minute intervals. Record your data.

2. Ask students to answer the two thought questions on the lab sheet while they are timing the water.

Laboratory III: What Type of Soil Do We Have?

Objective:
This lab will acquaint students with the soil types in their own region and the problems that can be associated with each.

Materials for each group:
- sticks and string for marking area
- garden spade or trowel
- 2 sandwich-size plastic bags for samples
- masking tape for labeling
- Centigrade and Fahrenheit thermometers
- metric rulers
- soil map from local agriculture extension agency

Procedure:
1. With the students, measure off a large outdoor area for each group and mark it with sticks and string. (Exact size will depend on the space you have available and the size of your class.)
2. Have each group use the top of their lab sheet to describe their area.
3. Ask groups to clear away about 10 cm² of space in the center of each area and record its air and surface temperatures in C° and F°. Each group will dig a hole and record the temperature, appearance, texture, and moisture content of the soil at two levels.
4. When students return to the classroom, bringing soil samples in labelled plastic bags, ask the following questions:
- What types of soil did you find?
- According to your data, are all of these soils native to this region, or have soils from other regions been added? (These might be sand in the playground or topsoil in the flower beds.)
- Look at the soil map. Do your findings agree with those of the soil service for your part of the country?
5. You may want to use slides from the local extension service to stimulate further discussion.

Earth at Hand

Laboratory IV: Who Else Uses the Soil?

Objective:
The students will observe and catalog the types of grass, flowers, and trees and the kinds of insects and spiders that thrive in their experimental area.

Materials for each group:
- plastic sandwich bag
- twist tie

Materials for the class:
- field guides or specimens to aid in identification
- large poster board for graphing and identifying plants

Procedure:
1. Instruct students to search out and observe the different types of animal life in their areas. They will record the name or a description for each, and count or estimate the number of specimens. To find earthworms, grubs, and so on, they may dig in the hole from Laboratory III, and then refill it.
2. Ask them to gather one leaf from each kind of plant in their area, and count or estimate the number of plants of each. (Caution students against picking plants: a single leaf will serve a whole group for identification.) Have them bring the plant specimens back to the classroom.
3. Indoors, students will use the guides to identify any creatures they did not recognize. Then they will contribute their data to a class graph of the animal and plant counts. Arrange the graph according to phylum, such as Insecta (insects), Arachnids (spiders), Annelids (earthworms), Bryophyta (mosses), Gymnosperms (conifers), and Angiosperms (flowering plants).
4. Use these questions to stimulate discussion:
- How do your group's findings differ from those of other groups?
- How do the air and soil temperatures relate to the types of plant and animal life you found?
- How does the soil type relate to the plant and animal life in your area?

Early spring is the perfect time to invite your students to get close to the earth. These laboratories will give them a solid groundwork for more advanced topics such as erosion, toxic waste, and the world soil orders.

Author's Note: Contact your local office of the USDA Soil Conservation Service or call your local Agriculture County Agent for valuable free resources on soil. Our County Agent comes into the classroom to work with the children. Maybe yours will too!

Data Sheet, Lab I

1. Rub some of each labeled sample between your thumb and forefinger. Record your observations.

	How does it feel?	How does it look?	Do the pieces break up?
SAND	_____	_____	_____
SILT	_____	_____	_____
CLAY	_____	_____	_____

2. Add water to a spoonful of each sample and squeeze.

	Crumbles	Sticks slightly	Molds a shape
SAND	_____	_____	_____
SILT	_____	_____	_____
CLAY	_____	_____	_____

3. Look at, feel, and squeeze the loam sample. It is:

_____Sandy loam (coarse-textured)
_____Silt loam (medium-textured)
_____Clay loam (fine-textured)

4. For extra credit: Put all three soil samples in a jar. Add water until the jar is 3/4 full. Shake well and let sit for 24 hours. Record your observations by drawing a picture.

National Science Teachers Association

Data Sheet, Lab III

1. Describe your group's area:

   • What is this soil used for? (Soccer? Grass? Flowers?)

   • How is the land graded? (Flat? Hilly? Slanted?)

2. Clear away a small patch of land (about 10 cm²) in the middle of your area. Take and record the following temperatures:

                                                   C°   F°

Air (Hold thermometer above the spot, shielding it from direct sunlight.)

Soil Surface (Place thermometer flat.)

At 3 cm (Cover thermometer with soil and wait 3 minutes to record.)

At 25 cm (Again bury thermometer and wait 3 minutes, uncover, record.)

3. Take soil samples at 3 cm and 25 cm, and describe:

   • How does the soil look? (Color, size of particles, other materials)

At 3 cm _____

At 25 cm _____

   • How would you describe its texture?

At 3 cm_____

At 25 cm_____

   • Which of these moisture ratings best applies?

    a. Dry: falls apart and sifts between fingers
    b. Slightly moist: appears moist, but does not stick together when squeezed
    c. Moist: sticks together in a clump when squeezed
    d. Very moist: water is visible when clump is squeezed
    e. Wet: water drips when clump is squeezed.

At 3 cm_____

At 25 cm_____

Data Sheet, Lab II

1. How long does it take the first drop of water to hit the plastic cup?

SAND _____ seconds

LOAM _____ seconds

CLAY _____ seconds

2. How much water is in the cup after 2 minutes?

SAND _____ mL

LOAM _____ mL

CLAY _____ mL

3. Stop, think, and write:

   • Which type of soil would you prefer if you were a plant? Why?
   • If your town were built on clay soil, what would happen during a hard rain?

4. How much water is in the cup after 5 minutes?

SAND _____ mL

LOAM _____ mL

CLAY _____ mL

MARGARET GOTSCH
Clear Creek School District
Houston, Texas
And
SHANNON HARRIS, Student
Friendswood High School
Friendswood, Texas

Sand isn't just for sandboxes and summer vacations. That gritty substance can be the stuff of fascinating lessons, as you introduce your students to arenology—the science of sand. Students might even learn to question some things they thought they knew: "white, silver sand," for example, is the last thing to look for on an idyllic, tropical isle, Hollywood fantasies notwithstanding.

To set up your own classroom sand laboratory, supply your students with several samples of different types of sand, from islands, the edges of continents, and the interiors of continents. If you seal a small quantity (25 cm^3) of each sample in a petri dish it may be reused many times. Students can examine the samples under a microscope at 10x or 20x magnification, looking for patterns and making some preliminary generalizations.

You can help your students interpret what they see under the microscope. If the sand grains are all about the same size, that is, have been geologically well sorted, they have probably been transported far from their original source. If they are well rounded and polished they were probably transported by water and/or worked over by waves for a long time. If their surfaces show frosting and pitting they were probably carried and/or abraded by wind.

With a little more information, students can attempt to decide whether their samples are from islands, continental shores, or inland beaches.

Island Sands

Most islands are either volcanic or biologic in origin. We should, therefore, expect that their beaches would be derived from basaltic (lava) or calcareous (seashell) materials.

The black beaches of Hawaii, Tahiti, and the Canary Islands are basaltic. The sand grains are usually polished obsidian or basalt from nearby volcanoes.

Most calcareous beaches, such as those at Nassau in the Bahamas, are composed of dull, chalky fragments (Figure 1). In some places shell material is so predominant that the sand almost resembles coquina, the soft, whitish limestone used as a building material. The beach at West Molokai in Hawaii is covered with vast numbers of small molluscan shells (Figure 2). The region's shallow, offshore waters provide prime breeding conditions for a variety of shellfish.

Continental Sands

Since continents are primarily granitic, we should expect that their beaches would be composed of materials derived from granite. The most resistant, common granitic mineral is quartz.

SANDY SCIENCE

Figure 1

Figure 2

Figure 3

Figure 4

44

Because travelling polishes the grains, beaches far from their granitic source will be almost wholly composed of fine, well rounded, well sorted, polished quartz grains. The beach at Ft. Myers, Florida is typical of this, and really could be called "white, silver sand" (Figure 3). Continental margins tend to be lined with quartz beaches.

Inland Sands

When we observe samples with large amounts of less resistant materials, such as the feldspars and dark minerals derived from granite, we may assume they come from the interior of the continent. One exception is beach material whose source is fairly nearby. New England's beaches tend to contain some feldspar and dark mineral material. The Mid-Atlantic states will show some mica and magnetite. However, in both cases the percentage of non-quartz mineral matter is very small. It seldom comprises more than five or ten percent of the total volume.

Rivers, lakes, and inland seas have beaches which cannot readily be classified. This is as true of Mediterranean sands as those around the Great Lakes. They exhibit a broad spectrum of sand types, from frosted gypsum crystals at White Sands, New Mexico to spherical limestone grains such as those found at the Red Sea.

River and lake sands tend to be of granitic materials with varying amounts of local minerals and rock fragments. The sands of Lake Superior contain large amounts of magnetite. The beach at Presque Isle State Park near Erie, Pennsylvania is shingled with flat fragments of shale and limestone (Figure 4). The river sand at Johnstown, Pennsylvania is peppered with bituminous coal fragments. Garnet, ruby, sapphire, and chromite have been panned out of stream sands from the Carolinas to Ontario, wherever pegmatites (coarse-grained, granitic, igneous rocks) have been intruded.

These examples are typical of the wide diversity found in the sands of the continental interiors. Your students should arrive at the conclusion that these sands cannot be categorically generalized in the same manner that islandic and continental margin sands can be. We can only say that they will reflect local geology and climate.

Data Tables

Make available at least five samples of sand from each of the main categories, and invite your students to arrive at a classification scheme which reflects the points outlined above. The table should include spaces to note the basic composition, impurities, sorting, transportation, shape, surface characteristics, and unusual properties of each sample.

Sources of Sand

Acquiring a wide range of sand samples won't be as difficult as you might imagine. Put a request on the faculty bulletin board, in parent newsletters, and on the Extra Credit board for students. Include a brief description of how you will use the sand, and a clip-and-copy label to be sure you get all the information you need. So far I've received sixty-eight samples, from as far away as Australia, China, Japan, Tanzania, Turkey, Saudi Arabia, Yugoslavia, and Normandy Beach. Most popular vacation resorts and islands are represented in the collection.

Your students travel. Your colleagues travel. They will enjoy this new perspective on the beaches they visit, and will return with bulging (and hopefully, neatly labeled) baggies of sand. As your students work with the samples in your class collection, they will take pride in having progressed from sandbox play to scientific arenology.

ALFRED C. PALMER
Penncrest High School
Media, Pennsylvania

Sample Label

Sand sample for [school name]: _____

Location: _____

Date collected: _____

Collector: _____

Comments: _____

(Please comment briefly on the source of these beach materials, and on the water, currents, prevailing winds, and other factors that might be of interest. We would also welcome your photos of the region.)

The water you showered or bathed in this morning might contain the same water molecules that swept over Niagara Falls a hundred years ago. And they might be the same water molecules that gently swirled over a coral reef in the South Pacific thousands, even millions of years ago. Sound impossible? Maybe not. Water has been called our most recyclable resource.

The distribution of the earth's total supply of water changes in time and by area. But the amount of total water remains basically unchanged. Water distribution changes according to a phenomenon known as the hydrologic cycle which is kept in perpetual motion by solar energy and gravity. Since this is a closed cycle, students can start to study it at any point.

Water and soil are natural resources in a delicately balanced interaction. Unfortunately, the intervention of humans sometimes adversely alters the balance of nature. The result may be water racing out of control, stripping soil from fertile ground causing widespread damage in human, economic, and environmental terms. Thus, it is important for students to be aware of the water cycle and its impact on the balance of natural resources.

Students can learn about the water cycle by studying a flow chart on the series of possible pathways which could be taken by a hypothetical water molecule. Each stop on the flow chart illustrated focuses on a section of the water cycle and the accompanying text gives background information and activities. The sections can be enlarged and arranged on a bulletin board, used as steps for a game, or as guidelines for a booklet. You can require as many or as few pathways to be taken as is appropriate for your time and curriculum.

Pathway one begins as a water molecule absorbs energy, changes from the liquid to the gaseous state, and leaves the ocean. From the gaseous state, the water molecule follows a path involving precipitation, ground water, cave formation, and/or living things, and an eventual return to the ocean where the cycle begins its repetition.

KENNETH USLABAR
Shattuck Junior High School
Neenah, Wisconsin

NOTE: Introductory copy and SCS information were provided by PAUL G. DUMONT, Soil Conservation Service, U.S. Department of Agriculture.

Pathway 1: OCEAN

The water molecule absorbs energy as the sun heats the ocean's surface. It begins to move progressively faster and escapes from the liquid state to become a molecule of water vapor.

Background
- The ocean contains 97 percent of the earth's water.
- Oceans cover approximately 75 percent of the earth's surface.
- A water molecule consists of two atoms of hydrogen and one atom of oxygen (H_2O).
- The average salinity of ocean water is 3.5 percent or 35 g salt per liter of water.

Activities
- Draw a water molecule.
- Define evaporation.
- On a world map, identify the world's oceans.
- List the solids dissolved in ocean water.

Proceed to Pathway 2.

Pathway 2: EVAPORATION

The water vapor molecule is carried aloft by air currents while prevailing winds move it across the surface of the earth. Evaporation leaves behind accumulated minerals the water molecule may have dissolved.

Background
- The change from liquid to gas is called evaporation.
- Evaporation contributes to the salinity of the remaining ocean water.
- One percent of fresh water is in the atmosphere.

Activities
- Draw the prevailing winds on a world map.
- Define vapor and its role in the water cycle.
- Evaporate one liter of sea water. Weigh what remains.

Proceed to Pathway 3.

Pathway 3: PRECIPITATION

The water molecule is carried so high that low temperatures cause condensation to take place. The water molecule joins with other water molecules to form a droplet. Many droplets join to form a drop which becomes too heavy to remain aloft. Gravity pulls the water molecule to the surface of the earth.

Background
- Rain, hail, sleet, snow, and drizzle are forms of precipitation.
- The greatest precipitation takes place in areas of greatest evaporation.

Activities
- Distinguish among precipitation, dew point, and condensation.
- Construct a rainfall map of the world.
- Write a report about acid rain.
- Compare rain, snow, hail, sleet, and drizzle. How are they alike, and how are they different?

Proceed to Pathways 1 or 4.

Pathway 4: LAND

A water molecule, falling as a liquid or solid, impacts the earth's surface with enough force to physically cause weathering. As the water molecule moves across the surface, it dissolves or carries away earth materials.

Background
- Differences in topography can determine precipitation patterns.
- Water is the universal solvent.

Activities
- Locate desert and rain forest areas on a map.
- Explain how rainfall determines housing, clothing, and lifestyles for people.
- Describe how land surfaces can be protected from the effects of falling water.
- Survey your community to find out about erosion control strategies.

Proceed to Pathways 5 or 6.

Pathway 5: SURFACE WATER

The water molecule flows across the ground dissolving minerals out of the soil and rock until it comes to rest in a small pond. The water molecule is cleansed of accumulated wastes by organisms living in the pond water.

Background
- Most communities get their water from surface sources.
- Marshes, lakes, and ponds are temporary storage areas for surface water.
- Fourteen percent of fresh water is available for consumption.

LA...

PRECIPITATION 3

STREA...

EVAPORATION 2

OCEAN 1

Possible Pathways along

Pathway 10: TRANSPIRATION/EXCRETION

In the course of being used in the life processes of a living thing, the water molecule helps break apart large molecules in the digestive system, becomes part of amino acids, and constitutes the liquid part of blood. Later, it is excreted by the animal to begin yet another path of the water cycle.

Background
• Water is used by living things to remove wastes.
• Water is used to transport nutrients.
• Water is essential for every life process.

Activities
• Describe how water is used in the process of photosynthesis.
• Explain plant and animal adaptations in regions of abundant water and regions that are dry.
• Describe how water is recycled on earth.
• Explain what happens in transpiration and excretion.

Proceed to Pathway 1, 2, 5, or 6.

Pathway 9: LIVING THINGS

The water molecule is absorbed by the roots of a plant, used to carry nutrients up the stem of the plant, and then is incorporated into an animal when the animal eats the plant. The molecule takes part in various life activities of the animal, such as respiration, transportation of nutrients, or the removal of waste.

Background
• Living things are made up mostly of water.
• It takes about 840 liters of water to produce one liter of milk.
• It takes 567 liters of water to produce a loaf of bread.
• Water is necessary for all life processes.

Activities
• Write out the chemical equations for photosynthesis and respiration.
• Describe how living things control their water content.

Proceed to Pathway 5, 6, or 10.

Pathway 8: CAVES

The water molecule is carried by a stream into an opening in the ground called a sinkhole and enters a world of darkness. The sinkhole is gradually enlarged by the water as are the cracks and fissures of the underlying limestone. Eventually, underground rock is dissolved to form caves.

Background
• Solution caves are usually found in limestone.
• Caves are the homes of plants and animals which have adapted to a dark environment.
• *Karst* is a term used to describe a region of underground drainage.

Activities
• Locate Mammoth, Carlsbad caves, and Luray Caverns on a map.
• Draw a cave, including a sinkhole, stalactites, stalagmites, and limestone layering.
• Interview a spelunker.

Proceed to Pathways 2, 7, or 9.

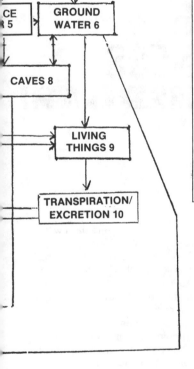

Pathway 7: STREAMS

The water molecule moves with many other water molecules in a stream bed. The molecule tumbles over waterfalls and meanders along valley floors. The force of the moving water molecules enables a load of dissolved material and material in suspension to be carried along.

Background
• Streams always flow downhill.
• Running water takes the path of least resistance.
• Less than 15 percent of the earth's fresh water is available for use.

Activities
• Locate and label major rivers on a map of the United States.
• Show where major cities are associated with rivers.
• Draw a profile of a stream.
• Identify stream related features in aerial photographs.

Proceed to Pathways 1, 2, 8, or 9.

Pathway 6: GROUND WATER

The water molecule infiltrates the soil until it reaches the ground water table at the top of the saturated zone. Various minerals, dissolved as mineral ions, attach themselves to the water molecule.

Background
• Almost all areas have ground water.
• Ground water usually moves only a few miles from the point of recharge to the point of discharge.
• Pore spaces below the ground water table are filled with water that generally moves downhill.
• Ground water can be polluted and is difficult to clean.
• Eighty-one percent of fresh water is in the ground and cannot be used.

Activities
• List the minerals which are most easily dissolved by water.
• Draw an underground picture, label the water table, show the saturated zone, a well, a swamp, and a lake.
• Describe how a soap solution may be used to test ground water for hardness.
• Find out about problems related to well water.
• Demonstrate water testing and water purification techniques.

Proceed to Pathways 1, 2, 5, 8, or 9.

Activities
• Define weathering, erosion, and solvent.
• Describe the operations in a water treatment plant.
• Describe the operations in a sewage treatment plant.
• Find out about flood control and build a model to demonstrate flood control techniques.

Proceed to Pathways 2, 6, 7, 8, or 9.

e Water Cycle

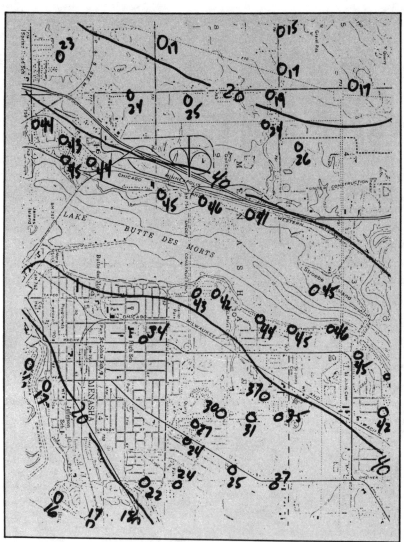

MAPPING
WATER
HARDNESS

If you live in a part of the country where most of the drinking water comes from wells, you can give your students a lab experience with close-to-home applications. Ask class members who have private wells to bring in a sample of the water in a small jar. Make sure the water has not been through a softener. Students who obtain their water from treatment plants can get a sample at the plant before it has been treated. Label each water sample with a location, such as the address of the student's home or the plant, and an identifying number.

Next provide a large map of the school district. Have students put the identifying numbers of their water samples on the map at the location where each sample was obtained.

Mix a water hardness testing solution by gradually adding granules of pure soap— *not* detergent—to distilled water in a dropper bottle. Shake the mixture, let it settle, and try it out in a test tube half full of distilled water. When five drops will produce a foam that lasts, you have arrived at the correct proportions. A test solution that has worked for me consists

of 100 g of soap powder added to and dissolved in 1,000 mL of distilled water.

Materials
- Map of your school district
- Liquid soap solution
- Small test tubes, 1.5 cm x 12 cm
- Test tube rack
- Droppers or squeeze bottles for solution
- Labeled water samples

Procedure
1. Fill a small test tube half full of the water to be tested.
2. Add liquid soap drop by drop, shaking after each addition, until the foam reaches a height of 2 cm and persists.
3. Repeat the procedure with a clean test tube. If results differ from the first, try the procedure again until you can duplicate your results.
4. Record the number of drops and the sample's number and location.

The number of drops that must be added to each sample to produce a lasting foam is the water hardness

number for that sample. Very hard water may require as many as 50 drops. Students record their numbers first on a class data table, and then on the school district map itself.

When all the water hardness numbers have been recorded, you and your students can draw lines on the map connecting points of equal water hardness. The pattern of lines will reveal regional variations in aquifer solubility and resultant water hardness. Certain aquifers are very soluble and produce very hard water, while other aquifers will have little effect on water hardness.

Whatever your results, this experiment will provide opportunities for discussion on the migration of water through bedrock, groundwater pollution, different kinds of aquifers, water softening processes, ion exchange, and other water quality topics of local and international significance.

KENNETH USLABAR
Shattuck Junior High School
Neenah, Wisconsin

The Water Planet An Optical Illusion

Student stands next to globe with 15 cm radius and holds small cubes that represent proportional amounts of ocean, frozen, ground, and fresh water.

Examine the globe and notice that over 70 percent of the Earth is covered with water. But appearances are deceptive in this case. The Earth is really a rock with thin films of water here and there. In fact, water is one of our most precious commodities.

The figures that follow show how you can build a model that will visually illustrate to your students how limited the Earth's supply of usable water is.

For those who like to play with numbers, the data in the tables offer unlimited possibilities. For example, have the students calculate how much fresh water per day each of the 5 billion people on Earth could consume before the supply would be drained. This calculation will help them appreciate the water cycle that is replenishing this supply. Also, they will have a better understanding of how water pollution threatens our well being.

MICHAEL B. LEYDEN
Eastern Illinois University
Charleston, Illinois

Reference
Amos, Fred C., "Earth Science Notes," *Ward's Bulletin.* Winter 1976, Vol. 16, No. 102.

TABLE 1
The Earth

Volume of a sphere:	V	=	$4/3 \, \pi \, r^3$ or $4.19 \, r^3$
Volume of Earth:	r	=	6400 km
	V	=	$1,100,000,000,000 \text{ km}^3$
Volume of globe with 15 cm radius:	V	=	$15,000 \text{ cm}^3$
Scale ratio of globe to Earth:	1 cm^3	=	$73,200,000 \text{ km}^3$

TABLE 2
Water—In Cubes

	Volume in Cubic Miles	Volume in Cubic kilometers	A Cube _ km on an Edge	Equivalent Volume On 15 cm globe Volume (cm³)	Cube Edge (mm)
Oceans	317,000,000	1,320,000,000	1100	18.1	26.3
Frozen Water	7,000,000	29,200,000	300	0.400	7.4
Ground Water	2,016,000	8,410,000	200	0.115	4.9
Inland Water	55,300	231,000	60	0.003	1.5

Conversion Factors:
4.17 cubic kilometers = 1 cubic mile

Volume of Earth is $15,000 \text{ cm}^3$ vs. $1,100,000,000,000 \text{ km}^3$ (rounded up from $1,098,100,000,000$ km^3 using 6400 km as radius of Earth)

This gives a ratio: 1 cm^3 represents $73,200,000 \text{ km}^3$ (rounded off from 73,204,415)

If you don't round off when computing the scale ratio, here's what you get: an Earth with a volume of $1,098,100,000,000$—and a globe that is $14,826 \ldots$ ratio is: 1 cm^3 represents $74,058,557 \text{ km}^3$. If you wish you can redo the scale . . . the differences are at the "tenths" or "hundredths" of mm when you make the model.

Elevate Your Students
With Topographic Maps

Rita K. Voltmer
Robert L. Paulson

Most courses in earth science use two-dimensional drawings to teach students to read a topographic map. Yet many students have difficulty visualizing land contours by studying lines on paper. Models can help. Constructing a contour model is not a new idea, but your students might find the following method helpful and enjoyable.

For each topographic model, you will need these materials:
- contour map
- corrugated cardboard, railroad board, mat board, or Masonite (any one of these may be used)
- opaque or overhead projector
- plywood or Masonite for the base of the model
- cutting tools (utility knife or Cut-Awl machine)
- white glue
- paint and/or modeling paste.

Selecting the site
Select a contour map of an area.

Rita K. Voltmer is an assistant professor of teaching/science, and Robert L. Paulson is an associate professor and director of instructional media at Malcolm Price Laboratory School, University of Northern Iowa, Cedar Falls, IA 50613.

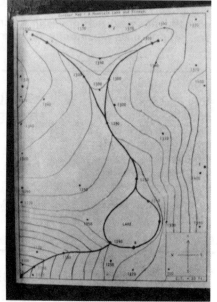

Figure 1　　　　*—Photos by Rita K. Voltmer*

Maps are available from several sources: earth science books, the U.S. Geological Survey,[1] as well as topographic maps made by teachers or students.[2] For our example in this article, we will use a map of a mountain stream and lake which was drawn by a teacher. (See Figure 1.)

Select a small area of the map for the model. We have found a workable size for the model to be 21 × 21 cm.

Obtain pieces of corrugated cardboard of uniform thickness. Grocery stores and appliance warehouses are good sources. Alternatively, mat board or railroad board can be used. Determine the approximate number of pieces of cardboard required by subtracting the lowest elevation on the map from the highest elevation and dividing this difference by the contour interval. For example, the lowest elevation in our example is 1100 ft.;[3] the highest is 1400 ft. The difference of 300 divided by the 20-ft. contour interval is 15. Therefore, about 15 to 16 layers of cardboard will be required for the model. Full sheets will not be needed for each layer. Scraps can be used for the higher elevations.

Drawing the lines
Place the topographic map on an opaque projector. Fasten a piece of cardboard to a bulletin board or the walls. (See Figure 2.) Focus the map on the cardboard, and trace the first two contour lines, beginning at the lowest elevation. Take down the cardboard; then put up another sheet of cardboard on the wall. Trace the second and third contour lines. Remove the cardboard and put up a new piece. Draw the third and fourth contours. Repeat the procedure until you have drawn all the contours.

[1] U.S. Geological Survey, 101 National Center, Reston, VA 22092.

[2] Printed materials for students on topographic mapping may be purchased from D. Louis Finsand, Malcolm Price Laboratory School, University of Northern Iowa, Cedar Falls, IA 50613.

[3] Since map scales are commonly given in feet and inches, customary units are retained here.

The actual map can be used with an opaque projector. But if you are using an overhead projector, transfer the original map onto a transparency. (If the map is printed with colored lines, be sure to make a photocopy of the map first.) The advantage of using the overhead projector is that outlining may be done in the classroom without having to dim the lights.

Figure 2

Taking shape

Cut out the contour lines of the model individually in one of two ways. Either cut the cardboard on a flat surface with an X-acto utility knife, or use a Cut-Awl (see Figure 3), an electric tool with a high speed vertical cutting blade. Adjust the blade length of the Cut-Awl so that it extends slightly beyond the thickness of the material being cut.

Remember to put another sheet of cardboard on the flat surface to avoid cutting the tabletop and dulling the knife blade.

Cut out all the cardboard pieces before building the model. This will allow you to make corrections or changes if you have made any errors. Mark each cardboard layer with its elevation and location. The outer pencil line is the cutting guide; the inner pencil line is the guide for placing the next higher contour level on the model.

Figure 3

Mount the lowest elevation cardboard layer with white glue onto a board of plywood or Masonite. After the glue has set, attach the next layer where the pencil line indicates. Repeat this procedure until all the layers are firmly attached.

You can enhance the appearance of the topographic model by painting it or filling in the contour intervals with some modeling paste. Our students label the elevations using pin "flags." (See Figure 4.)

All in all, the model has proven to be a valuable teaching tool for turning topographic maps into "real" contours of land. ∎

Figure 4

When Is the World Like

—————— **Verne N. Rockcastle** ——————

A few weeks before Halloween, a school district called to ask if I would visit one of their schools and "do" a science lesson as part of their general upgrading of the science program. I suggested Halloween week since that was convenient for me. But the school protested, saying that it was not a good time for them because, you know, the students and the teachers would be involved in all the Halloween festivities.

I asked myself why Halloween should interfere with a school's science program. It doesn't interfere with math or language arts. It doesn't interfere with the "important" things. Why science?

So I decided to develop a unit that included apples, pumpkins, and a Halloween party—but not at the expense of science. The following unit is adapted from the Halloween lesson I did go and teach. It is designed for the week before Halloween, and it shows how you can use some themes and props of the holiday to help students understand the ways we use longitude and latitude to find our bearings on this planet.

Lesson 1: Where in the World?

In this lesson students will learn how we set up the frames of reference that allow us to map our world.

To begin the discussion, divide students into small groups and give each group a ruler, a piece of paper, and a pencil. Put a dot (point) on a sheet of paper exactly like the ones you handed out, and show it to the class. Now, see if your stu-

dents can answer these questions:

- If I told you to make a dot on your sheet of paper so that it would be in the same position as this one, how would you do it? (Measure the distance from each edge.)
- Suppose the paper was so large that it didn't have any edges—it extended in each direction forever. Could you tell where to put your dot so that it would be in the same spot as mine? (No, because there wouldn't be any place from which to measure.)

Summarize by explaining that edges, boundaries, lines, and landmarks form *frames of reference* that help us locate points relative to them.

Complicate the problem by introducing a sphere. Stick a thumbtack into one of several identical, unmarked polystyrene foam balls. (See Figure 1.) Keep the ball with the thumbtack, distribute the rest of the balls to the class, and ask the following questions:

- Imagine that we are on a huge, unmarked ball out in space, where there is no up or down, no top or bottom, and no edges or lines of reference. And suppose this thumbtack represents my position on that ball. You are somewhere else on the ball where we cannot see each other. How can I describe to you where I am? How can you guide me to where you are? (There isn't any way because there is no top or bottom, no front or back, and no markings to serve as a frame of reference.)
- Now imagine that this gigantic ball is spinning. Would there be any point or points to serve as a frame of reference? (Yes, the top and bottom points of the axis about which the ball spins.)
- If the ball seemed to be spinning clockwise when you looked at one of those points on the ball, how would it seem to be spinning when you looked at

National Science Teachers Association

'umpkin?

the point on the other end? (Counterclockwise. Students may have a hard time with this idea. To clarify it, ask them to imagine that the ball has been flattened into a doughnut with spokes. It is now a bicycle tire, which is attached, of course, to a bicycle. A

This lesson shows that Halloween apples can be used for more than bobbing.

---

group of students standing on the right side of the bicycle as it moves to the right will see the wheel going clockwise, but a group of students on the other side, watching the same movement, will see the wheel going counterclockwise. The principle involved with a rotating sphere seen from the top and the bottom is the same.)

Before you go on to discuss how the top and bottom points of the axis can establish a frame of reference, remind students that the term for the opposite ends of an axis, like the one around which our planet revolves, is *pole*. Tell students to keep these two points, or poles, in mind as they think about the following question:

• How can we describe the thumbtack's position on the polystyrene foam ball? (By measuring the distance of the tack from the clockwise pole and from the counterclockwise pole, or vice versa.)

Explain that the poles of a spinning sphere are a frame of reference like the edge of the paper. Label the pole where the ball appears to turn counterclockwise *N* (for *north*) and the pole where

the ball appears to turn clockwise *S* (for *south*). And question students again about the thumbtack:

• Can you give the exact location of the thumbtack? (Not really. All that can be done is to describe the ring around the ball on which the tack is located. The point occupied by the tack could be anywhere on the ring: we need another frame of reference. [See Figure 2.])

Suggest that students form another frame of reference by designating a line connecting the north and south poles. Explain that this is an *arbitrary* line and that any other line connecting the north and south poles would do equally well. With this line and the two poles, students can describe the location of the thumbtack more precisely. But to make measuring the distance from each pole easier and more exact, have students designate another point of reference—a line halfway between the poles, which they can call the *equator* after the hypothetical line that circles the Earth midway between the north and south poles. (See Figure 3.) With these landmarks, students can state that the thumbtack is a certain distance N or S of the equator and a certain distance to the right or left of the designated line connecting the north and south poles.

You are now ready to explain latitude and longitude and how they are determined. Get out a globe and have students locate the north and south poles and the equator. Then point out the Greenwich Prime Meridian and explain that this is the designated line of reference connecting the two poles.

At this point, review the concept of degree and the number of degrees in a circle, a half-circle, and a quarter-circle. Explain that from the designated N–S line to another line halfway around the sphere is 180°. One could describe the distance all the way around the equator from the arbitrary N–S line as far as 360° by going either right (east) or left (west), but it is customary to use east (E) for 0° to 180° to the right of the N–S line and west (W) for 0° to 180° to the left of that line. (From here on, the designated N–S line will be referred to as the *prime meridian*, which is what it would be on Earth.)

Let students look at the globe once more and observe that it, too, is divided into 180° to the right and left of the prime meridian. Discuss and define the terms *latitude* and *longitude*.

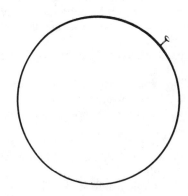

Figure 1. On a featureless polystyrene foam ball, there is no way of finding the location of a thumbtack stuck into the ball.

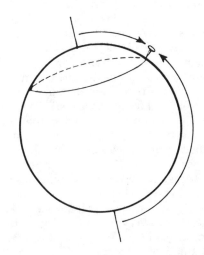

Figure 2. The axes of a spinning polystyrene foam ball make it possible to locate the thumbtack in relation to the two poles.

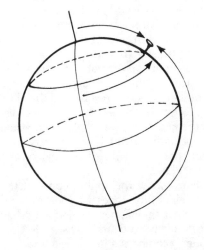

Figure 3. If a "prime meridian" and an "equator" are added to the foam ball, the thumbtack can be located precisely.

Before wrapping up the lesson for the day, question students further to make sure they understand the relationship between their arbitrary line and the prime meridian:

- Why is there a N–S line on which people agree?
- What would happen if each country had its own prime meridian?
- Was the prime meridian originally arbitrary?

Lesson 2: The Latitude and Longitude of an Apple

As this lesson will show, Halloween apples don't have to be used just for bobbing: they can also be used for plotting—in this case meridians, time zones, even your city or town.

Pass out a nearly spherical apple and a toothpick to each student. Have the students score an "equator" around their apples halfway between the stem and the remains of the calyx. Then have them score an arbitrary (what does the word mean?) prime meridian from the stem to the remains of the calyx.

Next, call for another meridian halfway (180°) around the apple from the prime meridian. (It might also help if students first drew a circle on a sheet of paper, marked 30° sectors on that circle to form a template, and put their apple at the center.) Then, have students prick lightly, in order, the following lines: meridians every 30° W and E of the prime meridian and parallels every 30° N and S of the equator. They should score these meridians and parallels more lightly than they scored the prime meridian so they can distinguish between the two designated lines of reference and the lines based on them.

Now, give the latitude and longitude for your own city or town and have each student stick the toothpick into the apple at approximately that position. Let the students compare apples to see whether each one has located the place correctly. If there are legitimate disagreements, explain that this method of locating a point is still approximate and that they will learn some refinements later.

Finally, use the apple to present the concept of time zones. Begin by explaining what time zones are and why they exist. Help students envisage a time zone by comparing its shape to that of a rather irregular orange section. In addition, tell them that the Earth has 23 full time zones and 2 half time zones and

Students cut the Eastern time zone out of an apple on which the seven time zones relevant for the United States have been marked. The apple rests on a paper template that was used to mark it accurately.

that a full zone covers 15° of latitude. After you describe the time zones, have students find and score the approximate location of the Eastern time zone, centered on 75° W.* When they complete the scoring, pass out paring knives, and let each student cut out this zone and eat it. (See photograph above.) Remind students how to use a knife safely, and give them cardboard and paper towels to protect the desks during the cutting portion of the activity.

Next demonstrate where the time zones listed below are on the globe, and have the students cut these zones out of their apples. (Each meridian listed below is the center of the time zone named with it.)

    Central (90° W)
    Mountain (105° W)
    Pacific (120° W)
    Yukon (135° W)
    Alaska-Hawaii (150° W)
    Bering (165°W)

When students have finished cutting out the zones, they can once again devour "time"—and then the rest of their apples.

*For other activities that help explain Greenwich mean time, see the October S&C Science Calendar, pages 27–30.

Lesson 3: Plotting With a Pumpkin

In this lesson, students will use latitude and longitude to plot a jack-o'-lantern. Not only will they put what they have learned into practice, but they will also provide the classroom with some appropriate Halloween decorations.

Have enough pumpkins for each group of four to six students. (Again, try to get pumpkins that are nearly spherical.) Instruct each group to use a fine felt-tip pen to mark the equator, the prime meridian as it appears on the globe, and Greenwich, England (51½° N and 0° W.) Then have students mark the northern parallels, each 30° from one another, and one parallel 30° south of the equator, as well as the meridians occurring every 30°. When this is done, have them plot the following points on their pumpkins:

45° N, 15° W	45° N, 45° W	30° N, 30° W
45° N, 75° W	45° N, 105° W	30° N, 90° W
30° N, 60° W	15° N, 75° W	15° N, 45° W
15° S, 30° W	15° S, 90° W	22½° S, 60° W

Next have the students connect the three points on each line above and cut out the outlined portions to make the jack-o'-lantern's eyes, nose, and mouth. (See photograph opposite.) They can

National Science Teachers Association

complete the pumpkin by cutting around the parallel that is 75° N to make a removable lid. If you want the game to be more challenging—and the classroom decorations more varied—provide a different set of coordinates for each team.

Lesson 4: Some Minute Particulars

This lesson will show students how the degrees of latitude and longitude are divided into minutes and seconds and will give them practice using these points of reference as they locate their own houses.

Begin by finding the lines of latitude and longitude on a topographical map of your city or town.** At the corners of the map students will find the latitude and longitude in degrees, minutes, and (in most cases, where the minutes are not integral) seconds. These will probably be new units, so you may wish to explain them.

Tell students that if they want to find a point as small as a house or farm on the map, locating it to within a degree will probably not be exact enough. (A

**Topographical maps are available from the U.S. Geological Survey. For ordering information, see Ava Pugh's "Happy Halloween!" on page 12.

degree of latitude covers about 70 miles, or about 112 kilometers [k]. A degree of longitude at the equator covers about 70 miles, too, but gets smaller as the Earth curves toward the poles.) To make locating a point—and mapping—more

> *You may need to expend some imagination to challenge students, but the results will be worth the effort.*

precise, degrees on maps are divided into smaller parts called *minutes*, just as an hour is divided into smaller parts called *minutes*. There are 60 minutes to a degree of latitude and 60 minutes to a degree of longitude.

- If a degree of latitude is about 70 miles (or 112 k), how much is covered by a minute of latitude? (More than a mile, but less than 2 k.)

Continue the parallel between measuring time and space by pointing out that there are 60 seconds in a minute of an arc and 60 seconds in a minute of

latitude, and then discuss the following questions.

- About how much distance is there in a second of latitude? (About 88 feet, or 27 meters [m].)
- Since degrees of longitude span smaller and smaller distances as you go farther north and south of the equator, about how much would a second of longitude be at your school? (Considerably *less* than 88 feet, or 27 m, depending on how far north, or south, of the equator the school is.)

As a final challenge, ask students to try to find the location of their houses on the topographical map in degrees, minutes, and seconds N, and degrees, minutes, and seconds W.

The Halloween Party

Circumstances permitting, you could end the week's activities with a Halloween party. If your students have been successful with the map exercise, you might consider a game in which different teams, each given only a list of coordinates and a topographical map of their city or town, compete to locate a mystery site. You might make the exercise more Halloween-like by choosing some spooky local spot—a well-known deserted house or an old graveyard—as the unknown locale.

But whether or not you continue the activities of the week at the party, you can survey the plotted pumpkins and remind yourself that holidays can be exciting times for learning. You may need to expend some imagination to challenge students at a level that will excite them, but the results will be well worth your efforts.

Using a felt-tip pen, a student who has already plotted the eyes on a pumpkin calculates where the nose should go. (For an idea of how the completed pumpkin will look, turn back to the drawing on page 8.)

Resources

Burns, Marilyn. *This Book Is About Time.* Boston: Little, Brown, 1978.

Latitude and Longitude. 2nd ed. 14 min. Color. Encyclopedia Britannica Educational Corp., 425 N. Michigan Ave., Chicago, IL 60611. 1981. Rental: $16.

Tannenbaum, Beulah, and Myra Stillman. *Understanding Maps: Charting the Land, Sea, and Sky.* New York: Whittlesey House, 1957.

Verne N. Rockcastle is a professor of science and environmental education at Cornell University, Ithaca, New York. Photographs by the author. Diagrams by Darshan Bigelson.

The Cardboard Cave

Earth science is one of the most difficult subjects to teach because you can't send students down into the Earth to observe geological processes. I have developed the Cardboard Cave to teach students about geologic time, cave formations, and cave life. To prepare for the construction of the cave, divide your students into five groups. Each group thoroughly researches its topic until it understands the processes involved in the formation of its part of the cave.

Group 1 covers past and present cave life, plant and animal temperature, environmental conditions, and cave formation. Groups 2, 3, and 4 study cave formations: Group 2 covers flow stone, curtain, rim pools, and popcorn formations; Group 3 studies stalactites, stalagmites, and drip lines; and Group 4 is responsible for pits, domes, and soda straws. Group 5 researches caves in the United States and the history of cave exploration.

Give students several days to complete their research, and then ask them to give an oral presentation to the class. I show slides of local caves at this point as well.

Students then use the following materials to construct a cave out of five freezer boxes placed with the open ends connecting to form a tunnel (See Figure 1.)

- five freezer boxes or large appliance boxes
- a cardboard barrel the size of a large trash can
- 50 newspapers and a batch of flour paste to make paper-mache or plaster of paris
- grey, black, and off-white waterbase paint
- paint brushes
- buckets
- empty toilet paper, wax paper, or copier paper rolls
- soda straws
- an old bed sheet to be used for flow stone and curtain formations
- a stapler to staple the sheet to the cardboard
- scissors

Have your students make the cave creatures and formations from the paper mache or plaster of paris. Give direction as needed. Group 1 is responsible for the opening of the cave and cave life. They should put a bat, cave cricket, spider, and salamander in the first box. Group 2 is responsible for the flow stone, curtain, rim pool, and popcorn formations, which should appear in the second box. Group 3 makes the drip lines and stalactite and stalagmite formations for the third box. Group 4 constructs the pit, dome, and soda straw formations for the fourth box; and Group 5 creates the crawl space at the exit from the cave. Very often the construction process for a formation parallels the geological process; for example, the gradual drip and buildup of rock material in and then out of solution is replicated by plaster of paris. Finally, if there are caves in your area, plan a trip so that students can put together what they have learned. One feature of this project is that you can teach as students come up with questions during construction, enhancing your students learning process.

Rebecca McMahan and
Mary Lee Parker
Austin Peay State University,
Clarksville, TN

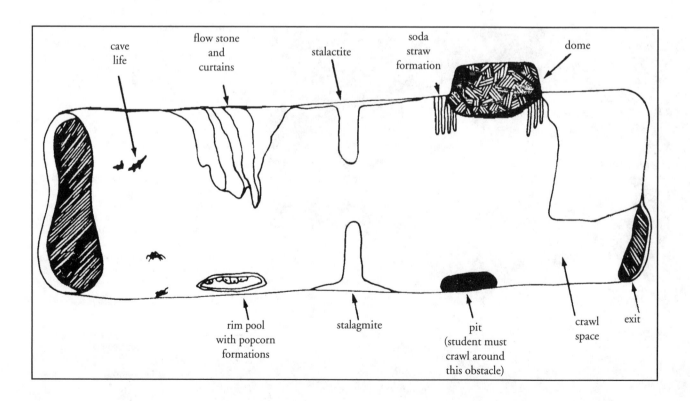

By Gregory Grambo

Orienteering, or, Which Way to Science Class?

A lesson in orienteering helps students figure out exactly where they're headed.

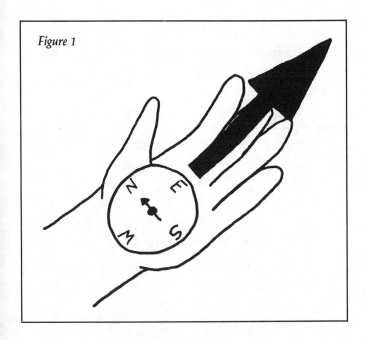

Figure 1

Long ago, explorers did not have the help of road signs to tell them which way to go. Instead, they relied on the North Star to calculate directions. They knew that when they moved toward it, they would be heading north, and they used that knowledge to develop the compass in the eleventh century.

Explain to your class that the face of a compass is divided into 360 sections, or degrees, like a pie that is cut into 360 pieces. No matter where you are, the compass will tell you what direction you are facing. Tell them that a compass contains a magnetized piece of metal that is balanced so that it can turn freely. The Earth's magnetic North Pole attracts the needle of the compass, causing it to point north. The opposite end of the needle, therefore, points south. If you hold a compass in your hand and the arrow points to your left, your body must be facing east (see fig. 1).

When teaching your students about orienteering, make sure they keep their elbows close to their bodies and their forearms stretched out in front of them with the compasses in the palms of their hands. Have each child align the compass by turning it so that the needle points to the N. The direction shown at the base of his fingers is the direction that the student is facing (see fig. 2).

Next, explain that *a pace* is the distance a person travels in one step. Show students how to walk paces. Ask them to explain how a pace is different from a meter. Make sure they understand that a pace is not an exact measure because it varies with leg length.

Figure 2

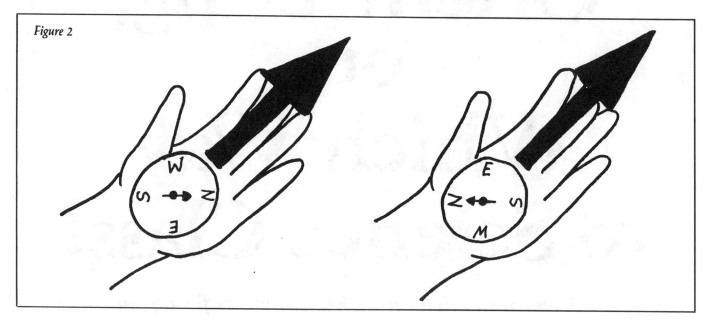

Figure 3, Reproducible Task Cards

A

1. Face 360°. Walk 30 paces.
2. Turn to 90°. Walk 30 paces.
3. Turn to 180°. Walk 30 paces.
4. Turn to 270°. Walk 30 paces.

C

1. Turn to 120°. Walk 30 paces.
2. Turn to 240°. Walk 30 paces.
3. Turn to 360°. Walk 30 paces.

B

1. Face 360°. Walk 30 paces.
2. Turn to 90°. Walk 70 paces.
3. Turn to 180°. Walk 30 paces.
4. Turn to 270°. Walk 70 paces.

D

1. Turn to 130°. Walk 30 paces.
2. Turn to 180°. Walk 15 paces.
3. Turn to 310°. Walk 30 paces.
4. Turn to 360°. Walk 15 paces.

After preparing your students for a lesson on orienteering, distribute task cards (see fig. 3) and a compass to each group of three or four students. Party-goods stores often sell inexpensive but accurate compasses appropriate for this activity. Have the children read the cards and predict the shape of the path each one describes. Then let them turn to the coordinates as directed on the cards and walk the designated paces, checking their predictions. Their results should match those given in figure 4.

To further sharpen their skills, send students on a treasure hunt. First hide an object in the classroom or in another part of the school. Then draw a map that students using a compass can follow to find the hidden object, and send them off to find the treasure!

For a homework assignment, have students map their neighborhoods. Ask them for compass directions to locate specific objects in their neighborhoods. Orienteering may lead students to explore further the exciting world of science. ॐ

Gregory Grambo is a science teacher at Louis Armstrong Middle School in East Elmhurst, New York. Illustrations courtesy of the author.

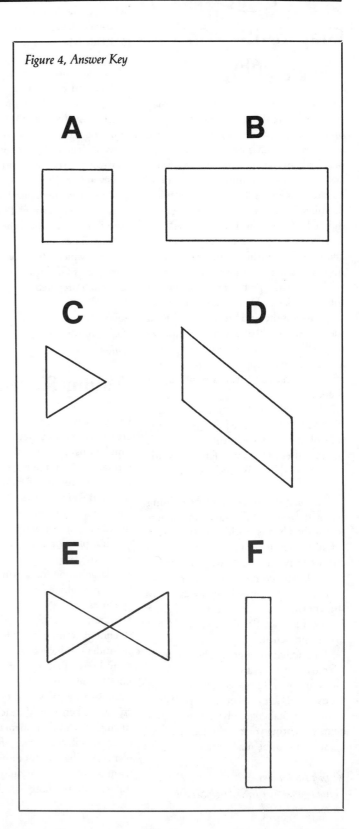

Figure 4, Answer Key

A B

C D

E F

E

1. Turn to 120°. Walk 30 paces.
2. Turn to 360°. Walk 15 paces.
3. Turn to 240°. Walk 30 paces.
4. Turn to 360°. Walk 15 paces.

F

1. Turn to 360°. Walk 80 paces.
2. Turn to 90°. Walk 10 paces.
3. Turn to 180°. Walk 80 paces.
4. Turn to 270°. Walk 10 paces.

Grapefruit Cartography

Two of the easier aspects of teaching map projections are showing the various types of projections and discussing the pros and cons of each. Unfortunately, the connection between the different projections and why they have limitations is not so apparent. Hence I developed a short but illuminating lesson in map projections dubbed, "Grapefruit Cartography."

After you get by all the wisecracks about fruit, sunshine, and home economics, assure your students that they really do need a grapefruit or orange for the next class. (Discourage the use of tangerines as the rind breaks easily.) Raid the kitchen for the rest of the supplies they will need: one sharp knife, spoons, and an ample supply of paper towels. With materials and fruit assembled, get them started.

Have students hollow out their grapefruit half and clean the juice off the outside of the rind so it can be marked. Instruct them to draw a familiar continent from the globe on the grapefruit skin.

The next step is the most entertaining part of the lesson. Tell your cartographers to take their hemispherical map and make it flat! It sounds simple enough, but the contortions this task will put them through are comic. Rips, tears, and wrinkles will be the order of the day. Any student who does not realize that making a curved surface flat distorts it, did not bring a grapefruit.

This activity is a good illustration to refer to while discussing projections. Grading is optional, as everyone can perform the exercise. One final note: keep a slab of wood and mallet handy for those students who bring you a wrinkled map and insist it cannot be made flat.

Kenneth Raisanen
Ontonagon, Area Jr/Sr High School
Ontonagon, MI

Review Game

I have successfully used this game which I call "New Jersey Geology—The Puzzle" for reviewing material in my Geology Concepts course; with minor modifications, it can be used in a number of disciplines and grade levels.

Cut two photocopied geologic maps of New Jersey (or an appropriate substitute) into ten pieces each and number them. Make four sets of corresponding numbered pieces of (plain) paper, and put two sets in each of two boxes labeled "team A" and "team B." You will have two boxes that each hold 20 pieces of paper—two 1s, two 2s, two 3s, and so on.

Divide the class into two teams (each with official spokesperson), and flip a coin to decide who goes first. Present a review question to team A, which then has 30 seconds to confer before their spokesperson gives the answer they have agreed upon (class notes and/or texts are, of course, not allowed). If team A answers correctly, pick a numbered piece of paper from team A's box—the goal is to be the first team to complete the (geologic) puzzle.

The fact that there are two sets of numbers in each teams box serves to make the game more exciting; if a number is picked for a puzzle piece they already have, no "reward" is given for the correct answer. Every third round, each team (upon supplying a correct answer) can opt to "raid" the other team and take a puzzle piece they need. If they answer incorrectly, however, their opponents can choose to answer the same question; if correct, they have a choice of either taking puzzle piece from the first team or going twice in a row. I have found that a team will often choose to "raid" the other team in order to avoid the possibility of having a duplicate number picked.

If a winner does not emerge by the end of the first review period, I record which team has which puzzle pieces and the game carries over to the second review session. "New Jersey Geology—The Puzzle" has proven to be fun and rewarding. The students come well prepared for the review, which is reflected in their exam grades.

Robert Metz
College of New Jersey
Union, NJ

Walnut Shell Geodes

Using walnut shells, your students can create classroom geodes—miniature simulations of the crystal-containing rock formations found in nature. This experiment calls for more student participation, but you will want to lend a guiding hand since students will be heating a mixture. You will need:

- a hot plate or other heat source,
- a kettle,
- a large spoon for stirring,
- copper sulfate (available at most drug stores),
- water,
- enough walnut shell halves for each student to have one,
- and several empty egg cartons.

Put 100 ml of water into the kettle and, while stirring, slowly add copper sulfate crystals. Then gently heat the mixture, stirring constantly. Students can take turns doing this if, for either safety or convenience, you have only one hot plate. Have them slowly and carefully add copper sulfate until no more will dissolve in the water. Remove the solution from the heat and let it cool slightly. Place the shell halves in the egg cartons, and carefully pour or spoon the cooled solution into the shells. Then set the cartons aside where they won't be disturbed for several days. When the water has evaporated, students can observe the crystal formations in the simulated geodes.

Caution: Copper sulfate is harmful if swallowed, so have your students wash their hands after handling the geodes, and don't let students take the geodes home if they have smaller brothers or sisters.

Students may have seen real geodes in nature or in a museum. Discuss how geodes form and bring in a sample if you can. Many small museums are willing to lend items to teachers for short periods of time.

Nona Whipple and Sherry Whitmore
Paul Culley Elementary School
Las Vegas, NV

STALAGMITES and STALACTITES

<u>What you will need</u>:
Box of Epsom salt (from local drugstore)
2 small jars
String (needs to be thick and water-absorbent)
Piece of wood board
1 large jar and a spoon
200 ml of water

<u>What you will do</u>:
1. Pour 200 ml of water in the large jar. Stir in Epsom salt to water until no more of the Epsom salt will dissolve.

2. Pour some of the mixture into both of the small jars.

3. Place the 2 small jars on the board. Put each end of the string into one of the jars. Let the center part of the string hang lower than the top of the jars.

4. Observe what happens over the next few days.

5. Which cone is the stalactite? Which is the stalagmite? How were they formed?

6. Where in nature could you find stalagmites and stalactites?

7. What would happen to these cones over a longer period of time?

Date With Science is a monthly feature developed by Lynne Kepler, Center for Science Education, Clarion University, Clarion, PA 16214.

Water, Weather, and the Environment

Disk-covering basic oceanography

By Mac Greenlee,
Monique Conway,
and Susan Miller

National Science Teachers Association

Today's Assignment: SECCHI DISK LAB

Art by Max-Karl Winkler

Our science department has always placed an emphasis on oceanography within the context of our study of Earth science. In a school year divided evenly between physical science and Earth science, we generally devote half of our Earth-science time to studying oceanography.

As a result of this concentration our department has found it necessary to develop some new labs to teach the subject. We cannot depend solely on textbook and AIMS (Activities that Integrate Mathematics and Science) labs to get us through the quarter.

With this in mind, we set out to develop several of our own investigations. Having learned about a few of the tools that oceanographers use, the teachers in our department decided to focus attention on the secchi (Sek-kee) disk. The secchi disk is a light-colored disk of variable diameter (generally six or seven inches) that is lowered into water to determine visibility, or water clearness.

Oceanographers use the secchi disk to take several visibility readings in a given area over a period of time. If there is significant change in the water clarity of a particular area, the oceanographers run tests on water samples to determine the cause of the change. Possible reasons for altered water clarity could include change in prevailing currents (for example, due to the presence of new sandbars), increased plant life (related to either natural or man-made events), or introduction of pollutants. Identification of the problem is the first step toward its solution.

The disk has a simplicity of function that won us over, resulting in the following lab activity. The apparatus for this lab is as basic as the secchi disk itself and can be easily assembled, disassembled, and transported from room to room. The lab serves a dual function: It introduces students to some of the basic concerns of oceanographers, and gives them hands-on experience with one of the most common tools of oceanography.

Materials
(There are many possibilities for substitution of materials on this list.)
- Five 1 to 1 1/4 inch rubber washers (You can find these in hardware stores; metal washers are also good for this activity)
- Aluminum foil
- Five 1 gram weights (fishing sinkers make a good substitute)
- One spool of string
- At least three long (30–50 cm), clear cylinders, wide enough to permit the secchi disk to pass through easily
- Ring stands and clamps
- Various suspensions—sand, silt, clay, and ash, for example—in water

Build a secchi disk

Wrap a small piece of aluminum foil around the rubber washer (leave the hole in the middle of the washer open) to provide a shiny surface that will reflect light.

Cut 50–60 cm of string from your spool (the length will depend on the height of the cylinders you use) and calibrate in 1 cm increments with indelible marker or pen. Tie the gram weight onto the string so that the string passes through the secchi disk and the disk sits on the weight (see Figure 1). If the secchi disk does not rest comfortably on the weight, you may need to select a larger

weight. (It is important that the disk remain nearly parallel with the floor while it is raised and lowered in the cylinder.) Make three or four backup secchi disks.

To duplicate various bodies of water, fill the three cylinders with separate mixtures of sand, silt, clay, or ash, and water. You may also want to combine materials, such as sand and clay. Experiment with different combinations.

Mix the resulting suspensions immediately before each group of students begins to take measurements. Cup your hand over the open end of the cylinder and shake gently until a satisfactory suspension is achieved. If your cylinders have drainage funnels, use water clamps to ensure that water doesn't leak out too rapidly. Slow leaks are easily handled by placing beakers under the cylinders.

Procedure

Divide the class into groups of three or four students each. Invite your first group up to the "disk-dunking" area. To ensure accurate readings, place the cylinders on the floor so that students can lower their secchi disks into the cylinders while looking directly down on the disks.

Instruct one student from the group to *slowly* lower the secchi disk into the first cylinder until the disk disappears from sight. At this point, the student stops lowering the disk and a second student determines where the water level reaches on the string. A third student records this figure in the group's lab notebook. The fourth student looks on during the first reading and then lowers the secchi disk for a second reading of the cylinder.

The members of the group continue to rotate tasks for the remaining measurements. Students should make three measurements of each of the three cylinders and enter the average of the three measurements for each cylinder in their lab notebook. (Groups of four students are optional—you may prefer other groupings.)

Class discussion

Ask students to answer the following questions based on their observations during this lab activity. (You may think of other questions that relate to your course of study.)
1. *Why do the cylinders have different depth readings for water clearness?* (Clarity varies according to the amount and size of the particles in suspension.)
2. *What is the relationship between size of particles and water clearness?* (Smaller particles remain in suspension longer, thus causing greater obscurity.)
3. *If the cylinders were left overnight, would the readings change? How?* (Yes, the readings would change. Even the finest particles will settle

to the bottom in time, yielding full-depth readings.)
4. *Outside of the classroom laboratory, what other factors might affect water visibility readings?* (Currents, plant life, pollution, and rain are some of the factors that affect water visibility readings.)

A word of warning

The secchi disk lab will not take a full class period, and it is necessary to plan a complementary lesson around it. Each lab group will need approximately five minutes to make nine measurements (three per cylinder). This means the activity will be completed in about 30 minutes, assuming that you use only three cylinders for the entire class.

Also, students will have a lot of free time while they wait their turn to use the lab apparatus. I usually distribute an AIMS worksheet or one that I have written myself for students to complete during this time.

It may be impossible for you to take your students to the ocean, but you *can* bring the ocean to your students. They are guaranteed to enjoy and learn much from their experience as classroom oceanographers. ▢

Resource

For more information on AIMS (Activities that Integrate Mathematics and Science), write to AIMS Education Foundation; P.O. Box 7766; Fresno, California; 93747.

Mac Greenlee teaches eighth-grade Earth science and middle level math at Sierra Middle School, Riverside, California.

Monique Conway is an eighth-grade physical science teacher at Sierra Middle School.

Susan Miller teaches seventh-grade life science and eighth-grade Earth science at Sierra Middle School.

The Significance of Rising Sea Levels

by Gregory J. Conway

*Why putting
your students
in hot water
may get the
rest of us out.*

Do your students know that the foam containers their fast food comes in could be warming up the Earth? Like carbon dioxide, the chlorofluorocarbons (CFCs) used in the manufacture of insulating foams are greenhouse gases. Last summer's drought brought the greenhouse gases and the warming trend of the Earth's climate into the public eye. The climate seems right for classroom discussions on this topic.

Visible proof

I have found, however, that unless students are presented with visual proof of the warming of the Earth, something they can actually see happening today, their enthusiasm for this topic runs out rather quickly. In order to combat this loss of interest, I came up with a warming-related activity that could be incorporated in a physical science, geology, earth science, or even an oceanography course.

The activity hinges on the fact that ocean levels increase as world temperatures rise and polar ice caps melt. An increase in tides accompanies the ocean rise. By graphing changes in tides and ocean levels over time, students can get a visual representation of how ocean levels have been changing.

During the activity, students will construct two graphs. In the first, students will graph sea level changes through the past 30 000 years. The graph should show what scientists have found through radiocarbon dating—that sea level was close to its present level 30 000 years ago, then dropped about 125 m 15 000 years ago, at the time of the last Ice Age.

The second graph will show tide changes for particular coastal cities. For these specific data, you either can utilize the information on New York City and Charleston in Figure 2 or have your students collect data for cities of their own choice from past issues of the *Old Farmer's Almanac*. The almanac includes tide information for most U.S. coastal cities. The average tidal change (plus or minus) from the mean low tide of the preceding year is given for each year and students could investigate as far back as possible. As an alternative, you could send for tide tables published annually by the National Ocean Survey (See Note).

You also will have to select a time frame for the second graph. The time frame for this activity can vary greatly depending on how you would like to approach it. Check and see how long your school library holds past issues of the yearly almanacs. You may want to bring a class in for a visit and have the students work in groups with the

—*Photo courtesy of Al Wolfe Associates*

National Science Teachers Association

Figure 1. Variations from present sea level for the past 30 000 years.

Variation (m)	Years from present (thousands)
0	0
-6	5
-61	10
-126	15
-55	20
-18	25
-15	30

Figure 2. Changes in sea level over 5-year intervals for New York City, N.Y., and Charleston, S.C., from 1900–1970.

Year	Changes in sea level (cm) New York	Changes in sea level (cm) Charleston
1900	7.0	0.0
1905	3.5	0.0
1910	4.5	0.0
1915	6.5	0.0
1920	8.0	0.0
1925	7.5	-3.0
1930	9.0	-1.5
1935	13.0	6.0
1940	14.0	3.0
1945	16.0	13.5
1950	19.0	11.6
1955	17.0	13.0
1960	17.5	13.5
1965	21.0	12.0
1970	23.0	15.0

various almanacs so they can get an idea of how scientists gather their information. You also might want to review with the students how to gather and keep track of their own information.

After the students construct their graphs, they should work on the following questions:

• What caused the dip in sea level from 10 000 to 20 000 years ago?
• What happened to the ocean water during the Ice Age?
• How much has sea level risen for New York City since 1940?
• What was the rate of sea level rise for New York City from 1940 to 1955? From 1955 to 1970?
• Discuss why the rate of sea level rise might be increasing.
• What impact does the sea level rise have on human activities?

• Using the rate you calculated for 1955 to 1970, calculate how long it would take for the sea level to rise 3 m and 6 m.

Remember that you might have to make some substitutions to take into account the selected cities and years under study. It should become apparent to students that not only is sea level increasing, but also that the rate of rise actually might be increasing as well.

A map of the world's oceans showing the widths of continental shelves would be appropriate in any follow-up discussion as it would help students see why coastal cities may vary with reference to sea level changes. The map will show that near some cities, the continental shelf is flat and extends far out from the continent. These cities suffer greater beach loss than cities in areas where the shelf is not as wide.

A good follow-up discussion also should cover the reasons for the rise in oceans. Your students may think the connection is a bit farfetched, but CFCs and carbon dioxide do contribute to the problem of rising sea levels. Of course, other factors contribute to this situation as well; tell your students about tectono-eustatic changes (which result from changes in the shape of ocean basins), glacial isostasy and eustasy (defined as the increase or decrease of sea level due to receding or expanding glaciers), and hydroisostatic deformation (which results when

---

Sea level has changed dramatically throughout the Earth's history.

---

an increase in the amount of water over the continental shelf actually causes it to sink).

Sea level has changed dramatically throughout the Earth's history and will continue to change. In our day, though, sea level rise is becoming more and more visible. Coastal highways and roads are inundated during extreme high tides more often than in the past; beaches are becoming more narrow. Anything which contributes to the warming of the Earth, including greenhouse gases such as CFCs and carbon dioxide, adds to the problem. By using this activity in conjunction with a discussion of greenhouse gases, the greenhouse effect, and the warming of the Earth, you may make your students interested in this problem. You also may contribute, in your own way, to clearing it up. ∎

Note

Information on tides is available from both the *Old Farmer's Almanac* (Yankee Publishing, Inc., Dublin, N.Y.) and from tide tables published by the National Ocean Survey (Distribution Division, 6501 Lafayette Ave., Riverdale, MD 20840)

Gregory J. Conway teaches lab biology and marine science at Highland Regional High School, Erial Rd., Blackwood, NJ 08012.

Slick Science

Oil spill education is an environmentally sound practice.

By Jane O. Howard

The news media does a good job of keeping the public informed of the oil spills that seem to occur on a regular basis (see sidebar, p. 21). But such spillage comprises only a small portion of the oil that is contaminating one of our greatest natural resources: water. In terms of this contamination and its subsequent damage to the environment, we may be our own worst enemies.

Unnoticed Spills

Oil spills are not necessarily accidental, nor dramatic. The majority of the oil in the ocean has little to do with the media-reported off-shore accidents and a lot to do with some unnoticed, everyday occurrences. Look at a few facts from the National Academy of Science's statistical analysis of petroleum entering global waters (1973):
- More than 54 percent of the oil in the ocean came from land-based (nonpoint) sources such as storm water run-off, leaks from storage facilities, and industrial processes.
- Marine sources, including tanker operations, contributed 18 percent, much of it due to the ordinary process of cleaning and flushing tanks at sea. Tanker accidents contributed 11 percent.
- Bilge discharges were responsible for 8 percent of the oil in the oceans.

Perhaps the most threatening damage comes from persistent low-level oil pollution, such as waste oil from cars and leaking gasoline tanks. At least 40 percent of all non-point oil pollution may result from motorists' changing their own oil and disposing of it improperly. Ideally, used motor oil should be taken to a service station, where it is then picked up by a waste management company to be recycled or burned. Proper disposal of used motor oil is important because oil picks up toxic contaminants and carcinogens, such as lead and zinc, during engine use; when this oil is emptied into landfills, storm drains, or backyards, it carries these toxicants to ground water, streams, and lakes.

What Happens to the Animals?

The effects of oil on the marine environment are complex. There are many types of oil compounds, and each type and the amount of it relates differently to such environmental factors as temperature, light, salinity, and weather. Depending upon growth stages and time of year, marine organisms may be dramatically affected by a spill or affected very little. For example, snails may lose their ability to attach themselves to a rock, and be swept away and killed. Fish may ingest oil and, depending on how the compounds react with their individual cells, clog their gills. Crustaceans and fish eggs may be affected by toxic materials found in even a thin layer of oil on the surface of the water.

If the oil sinks into the water's sandy sediment, or covers its beaches, some species (oysters, clams, mussels, smelt, and herring) may not spawn offspring. If they do, the offspring may be affected.

National Science Teachers Association

In some cases, however, fish may be resistant to the effects of oil in the water because of their protective mucous membrane. Certain types of marine life (flora on barnacles and some worms) can even benefit from the spill by attaching themselves to a floating oil lump.

Sea birds that skim the surface of oily water often coat their feathers and natural oils with contaminant oil, thus losing their ability to swim, fly, or maintain their body heat. If a bird cannot maintain its body heat, it dies, just one of the reasons why the number of wildlife casualties resulting from oil spills is so high.

How Do We Clean Up?

All cleanup methods present difficulties, and some can be as damaging to the environment as the spill itself.

Chemical surfactants, in the United States used only under strict regulations to control spills, act like detergents to dissipate oil. Though these dispersants contain toxic chemicals, they cause less damage than a large block of oily water would. This method also offers ease of implementation, as the surfactants can be dispensed from airplanes. Often dispersants are the most feasible solution to an oil spill, particularly in rough seas.

Mechanical cleanups, where oil is contained by booms

Fourth graders at the Poulsbo (Washington) Marine Science Center test various methods of cleanup with used motor oil and water.

and absorbent materials like straw, cotton, or nylon, are more expensive and take longer to implement than do dispersants.

Finally, biological methods focus on using hydrocarbon-consuming bacteria, or "oil-eating microbes," to process the hydrocarbons involved in the spill. Highly controversial, this method was used to help control the enormous spill resulting from the *Exxon Valdez* tanker accident.

Spilling into the Classroom

The following activity allows teachers an invaluable opportunity to encourage environmental awareness in the classroom. By combining facts and hands-on learning, students can develop the analytical skills necessary to deal with problems like oil pollution.

For each pair of students you will need
- one aluminum pie pan half-filled with water,
- a medicine dropper full of used motor oil,
- cotton balls,
- nylon,
- string,
- paper towels,
- liquid detergent,
- and feathers.

Give each pair of students the materials and a work sheet on which to record their observations. Ask students to make predictions about the action of oil and water.

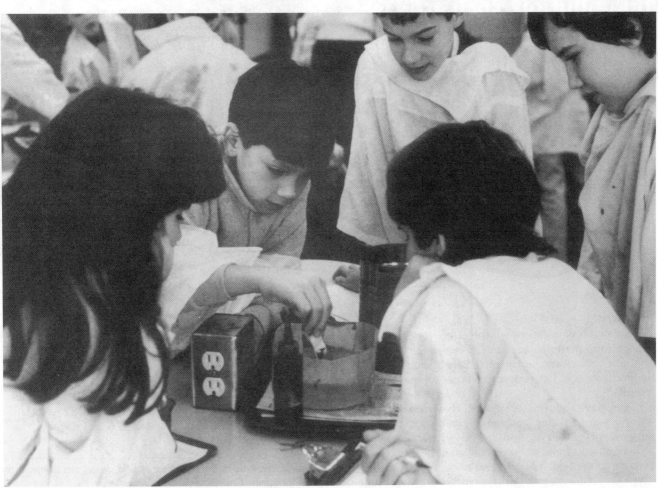

The Poulsbo Marine Science Center

What do you think will happen to the oil when you drop it on the water? Will it sink, float, or mix in? Which material do you think will clean up the most oil in the least amount of time? Cotton, nylon, paper towel, or string?

Have each pair create an oil spill by putting five drops of used motor oil in the "ocean" (aluminum pie pan). Let them observe the action of the oil and record what happens. Ask, Were your predictions correct?

Ask students to predict the effect of wind and wave action on oil and water. They can simulate the ocean's behavior by blowing on and moving the water in the pie

can try the procedure using fresh water and then salt water.

Conclusion

During this oil spill simulation, students gain a real-life understanding of what happens when an oil spill occurs. In doing so, they also develop a greater awareness of some of the difficult choices that must be made if we are to take responsibility for the maintenance of the world in which we live. Your students may not yet be changing their own motor oil or making policy, but starting early

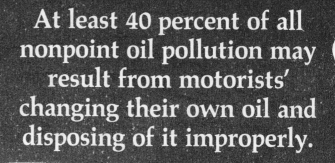

Perhaps the most threatening damage comes from persistent low-level oil pollution, such as waste oil from cars and leaking gasoline tanks.

At least 40 percent of all nonpoint oil pollution may result from motorists' changing their own oil and disposing of it improperly.

pan. What happens?

Have each pair determine the effectiveness of each of the cleanup materials provided. They should identify the amount of oil cleaned up by each material and how quickly it worked. Ask, Do your predictions compare with the results?

Then have them make another five-drop oil spill in a second pan of water, add five drops of liquid detergent (dispersant), observe, and record what happens. Ask, Where do you think the oil would go in the real ocean?

Let students dip a feather in the oil. Ask, How do you think oily feathers might affect birds' behavior? Students

with environmental education bodes well for the future—theirs and ours.

Resources

Hunter, R. (1987, November). *Oil and Hazardous Waste.* Presented at the meeting of the Northwest Association of Marine Educators. Poulsbo, WA.

Jane O. Howard is the supervisor of field studies and camps at the Pacific Science Center in Seattle, Washington.

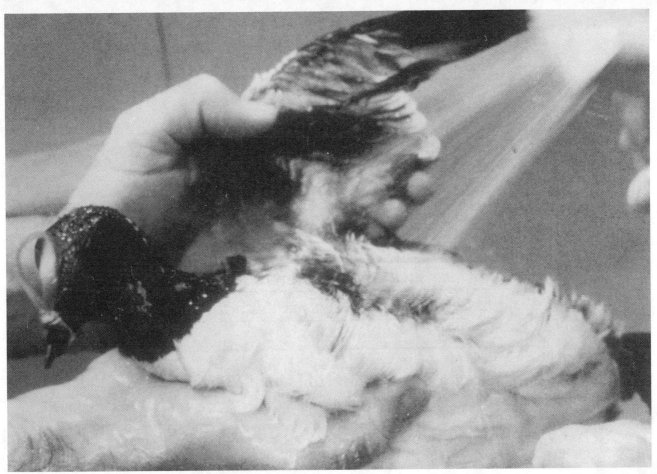

Disaster Spills: A Drop in the Bucket

March 18, 1967—Torrey Canyon Spill. A tanker crash entering the English Channel left 90,000 metric tons of oil washed up on the coastline of France.

December 15, 1976—The vessel *Argo Merchant* spilled 28,000 metric tons of oil (equivalent to 17,000 railroad tank cars) off Nantucket Island, Massachusetts.

January 28, 1981—Thirty-five hundred metric tons of oil spilled into Galveston Bay, Texas, from tanker *Olympic Glory*.

July 30, 1984—The *Alvenus* tanker (now the *Halifax*) spilled 9,100 metric tons of oil in Lake Charles, Louisiana.

October 31, 1984—Forty-seven metric tons of oil spilled from the *Puerto Rican*, less than 320 km west of the Golden Gate Bridge in San Francisco, California.

December 21, 1985—The *Arco Anchorage* ran aground, spilling 700 metric tons of oil into Pacific waters near Port Angeles, Washington.

December 28, 1988—Nearly 900 metric tons of oil was dumped from the barge *Nestucca* into Grays Harbor, Washington.

March 24, 1989—In a tanker accident, vessel *Exxon Valdez* spilled thirty-five thousand metric tons of crude oil into Prince William Sound in Valdez, Alaska.

These accidents and many others have found their way into newspaper headlines. Current event headlines such as these are an excellent introductory source to foster students' awareness of the environment. Teachers can further develop this awareness, however, by taking advantage of some of the newest classroom resources that incorporate these

Even with a thorough cleansing not all wildlife can be saved. Sea birds that preen their feathers can swallow enough oil to suffer direct effects.

current events, while also teaching students how to behave responsibly in the environment.

Additional Resources in Oil Spill Education

- *Alaska Fish and Game* (July-August 1989) produced a special issue devoted to oil spills and their cleanup. Copies are available for $2. Address requests to Editor, *Alaska Fish and Game*, Box 3–200, Juneau, AK 99802.

- FOR SEA Marine Science Curriculum Guides (K–12) contain a variety of activities and teacher information to bring environmental education into the classroom. Funded through the National Diffusion Network (a division of the United States Department of Education), FOR SEA staff offer inservice workshops across the country to help teachers implement marine science education into their existing curricula. For information about inservices in your area or how to obtain the curriculum guides, write FOR SEA, 17771 Fjord Drive, N.E., Poulsbo, WA 98370; tel. 206-779-5549.

- Lethcoe, N., and Nurnberger, L., Eds. (1989). *Prince William Sound environmental reader*. Valdez, Alaska: Prince William Sound Conservation Alliance. Retail $10. Available through the Prince William Sound Conservation Alliance, P.O. Box 1697, Valdez, AK 99686.

The Acid Rain Debate

Turn your students into industrialists, ecologists, and government officials in this acid rain role-play.

by Rodger Bybee, Mark Hibbs, and Eric Johnson

How can we explore with our students a topic as many-sided and controversial as acid rain? When even scientists and government officials are hard pressed to keep up with research findings about the environmental impact, industrial connections, and social and ecological implications of acid precipitation, how can we expect our students to understand the issue?

Last month, *The Science Teacher* offered an overview of the problem of acid rain in North America ("Acid Rain: What's the Forecast?" by Rodger Bybee, pp. 36–47). Having introduced your students to the issue, you can help them sort out their own opinions about acid precipitation, its sources, and its consequences with the simulation activity we have designed. So review the cast of characters, brush up on your acid rain facts, and prepare to make some heavy decisions about one issue that strikes at the heart of science/society questions—acid rain.

By the time this activity is over, your students should be able to describe the causes and the effects of acid rain to anyone who wants to know and to suggest several concrete solutions to this international problem.[1]

[1]This program was produced through the Science, Technology, and Public Policy Program at Carleton College and was partially funded by a grant from the Sloan Foundation.

Setting the scene

If you want to try this activity with your own class, you might find the following suggestions helpful. Allot half the first period to assign a role to each student. If you have 24 or fewer students, designate 4 to serve as members of the International Commission on Acid Rain. Let the rest of the students fill the shoes of as many of the other 20 characters as possible. (Be careful in assigning roles to avoid the stereotypes of woman secretary, male geologist, male company president, and so on.) If your class is larger, simply expand the size of the commission.

Explain to your students that the commission, which was set up by the United States and Canada, is going to conduct a public hearing. On the basis of the testimony it hears, the commission will make recommendations to both governments on specific ways to solve the acid rain problem.

The members of the commission then should select a chairperson and a secretary. The chairperson will be responsible for recognizing speakers, maintaining order, and speaking for the commission. The secretary must keep a record of each person testifying (name, occupation, and group he or she is representing) and summarize the views of each speaker.

Before the next class period, the other students should become familiar with the characters they are about to play and be ready to understand and speak each character's point of view. The following day, the hearing begins. Each witness is allowed to speak for 10 minutes. At the close of the hearing, the commission members meet in private to compile a list of their recommendations. The possibilities are varied: The governments should take no action at the present time, they should study the matter

further, they should impose certain pollution restrictions, and so on. In reaching a decision, the commission must be aware of the international ramifications of the acid rain problem. Commission members can use the questions listed on the role sheet that accompanies this article as one of their resources.

At the close of the period, the chairperson announces the recommendations to all those attending the hearing. After the hearing, reconvene as a class to discuss the implications of the commission's recommendations:

• Which testimonies might have been biased? Why do you think they were? Do you think all the evidence was biased?

• Can the commission make valuable and sound recommendations even if some of the testimony was slanted?

• On what criteria did the commission base its decisions?

• Was the commission fair to all interested parties? Should fairness be an issue?

• Did the commission weigh and then balance the costs and benefits that will arise from its decisions?

• Can the commission's recommendations realistically be implemented?

• Did the commission's recommendations address the international aspect of acid rain?

Your class might think of other activities to pursue related to acid rain. Some that have been suggested to us include the following: Write to individuals or agencies such as those portrayed in the simulation to find out their positions on acid rain. Write a report comparing current regulations and research findings to the recommendations of the International Commission. Write to the U.S. Environmental Protection Agency (William Ruckelshaus, Director, U.S. Environmental Protection Agency, RD-674, Washington, DC 20460) and the Ca-

nadian Ministry of the Environment (Honorable John Roberts, Minister of the Environment, Ottawa, Ontario, Canada K1A 0H3) requesting information on acid rain as well as on what actions each agency is taking.

Guidelines for the commission and descriptions of each character appear on the following pages. Photocopy and distribute what you need. It's not an easy business trying to help your students unravel a science/society issue as complex as acid rain. But at the close of your hearing, we think you'll find it's been worth it.

Continued on p. 52.

Rodger Bybee is an associate professor of education at Carleton College, Northfield, MN 55057. Mark Hibbs is an engineering student at Stanford University, Palo Alto, CA 94303, and Eric Johnson is a former biology teacher who lives at 613 Union, Northfield, MN 55057.

The Script

Guiding the commission

As members of the International Commission on Acid Rain, you are expected to be nonpartisan. You are to make recommendations to the Canadian and U.S. governments about how they should individually or jointly deal with the problem of acid rain. You will make these recommendations after conducting a public hearing where private citizens, representatives of industry, and government officials share their information, opinions, and suggestions. As commission members, you should consider the following issues:

• Does the United States have any responsibilities to Canada? Does Canada have any responsibilities to the United States?

• Does private industry have a responsibility to citizens, to the government, and, in particular, to foreign governments? Do citizens and governments have responsibilities to private industry?

• Can we allow acid rain to continue to fall if, for example, many of the fish and other animals in the Boundary Waters area of Minnesota and in similar regions begin to disappear? Can we afford to impose pollution control restrictions that may cost millions or billions of dollars and thousands of jobs?

• If the government mandates pollution control by industry, does the government have a responsibility to help pay for the necessary equipment, such as scrubbers?

• Does industry have a right to endanger our health or to threaten the environment in order to log a profit or to provide jobs?

• Who should be responsible for the costs of reducing acid rain? Industry? Government? Citizens?

• Who is responsible for supporting research on acid rain? Is more research necessary?

• Must we always completely understand such a problem before we can act on it?

• How should we take into account the fact that at times acid rain has some beneficial effects, such as providing nutrients to increase the yields of some crops? Should we try to reduce acid rain if it benefits one sector of society, is harmful to another, and is regarded neutrally by a third?

Playing the role

Hao Ling/Wen-Zhi Chen, age 28, geologist for the U.S. Geological Survey

From experience, you understand that unbuffered soils are sensitive to acid rain and that soil acidification can have far-reaching consequences. One is the leaching of metals such as aluminum and iron, which can harm plants. Plant nutrients may be lost from susceptible soils. Some soils, such as those where Canada's forestry industry thrives, are naturally acidic but sensitive to increases in acidity. Thus, while additional research under field conditions is clearly necessary, studies already indicate that acid rain threatens both aquatic and terrestrial systems.

Alex/Alexis Scott, age 37, chemist for Ontario Hydro, which provides electrical power for the Province of Ontario

Ontario Hydro is the second largest source of SO_2 within the province and the largest industrial source of NO_x. Ontario Hydro's coal-fired stations produce about 20 percent of the total emissions of these two pollutants in the province. Hydro is active in controlling SO_2 emissions by using only washed coal, which reduces sulfur levels by 15 to 20 percent. Hydro is also installing low NO_x burners and SO_2 scrubbers to combat acid rain. You would urge the commission to require similar standards for SO_2 and NO_x sources in the United States, since Hydro's actions alone are not effective because polluted air masses enter Canada from the United States.

76

Julia/Jason Kirk, age 36, consulting meteorologist based in Toronto, Ontario

As a member of a Canadian-U.S. research group, you have determined that the levels of transboundary pollution depend on prevailing wind and weather patterns. The net flow of sulfur is from south to north across the U.S.-Canadian border. On the average, three to four times as much sulfur crosses into Canada as moves into the United States, although the flow of sulfur from U.S. sources to Canadian air space is close to the total Canadian emissions in volume—5.5 million tons. You feel strongly that the commission should consider these data in making its recommendations.

Hannah/Jacob Frisch, age 62, officer with the Minnesota Department of Natural Resources (DNR)

Although DNR studies have not yet found an acidified Minnesota lake, one quarter of Minnesota's 12 000 lakes have been identified as susceptible to acid rain. Canadian studies show that in addition to the threat to fish, amphibians, including salamanders and frogs, are drastically affected by acid rain. Because the salamander plays an essential role in the food chain, its disappearance would negatively affect waterfowl, such as loons, and mammals, such as skunks and shrews, that eat salamanders. Therefore, you believe the commission should consider that acid rain threatens wildlife as much as fish, amphibians, and aquatic plants.

Leroy/Loretta Nelson, age 55, owner of Nelson's Lodge on the edge of Minnesota's Boundary Waters

Your lodge caters to fishing enthusiasts drawn to the lakes of the Boundary Waters. While a report to the legislative commission on Minnesota resources found that no lake in the state was acidic or acidified, Minnesota Department of Natural Resources officers have told you that because of soil types and granite bedrock in your area, the local lakes are sensitive to acid deposition. Moreover, highly prized walleye pike, smallmouth bass, and lake trout comprise a major portion of the fish in these susceptible waters. If these lakes become acidic, fish populations will probably decline; this in turn would reduce the number of resorts and outfitter operations in the area by at least one half. You are representing these businesses. Acidification of Boundary Water lakes could cut down the volume of your industry from $63 million to $21 million annually; lost jobs are estimated to be 3000. You will urge the commission to take immediate steps toward eliminating acid rain.

Lyle/Lucille Butler, age 51, farmer from Lakefield, Ontario

You live in an area that is reportedly threatened by U.S.-produced acid rain. Although you've heard that acid rain can damage crops such as spinach, let-

> ## On the basis of the testimony it hears, the commission will make recommendations to both governments on specific ways to solve the acid rain problem.

—Johanna Vogelsang

tuce, bush beans, and radishes, and can reduce yields of some crops such as beets, carrots, radishes, and broccoli, other crops have had higher yields thanks to the fertilizing effects of acid rain. In your case, corn yields have not declined, but occasionally have increased. You feel the commission should recognize that acid precipitation can sometimes be beneficial.

pressed economies of these regions.

Theresa/Paul Duchamps, age 33, environmental engineer for Provincial Mining Company, Manitoba, which produces nickel and manages several large mining and smelting operations
Your company would like the commission to consider recommending action to return acidified lakes to their nor-

systems apparently carry SO_2 particulates into eastern Canada where they are deposited in acid rain. Your industry is attacked for contributing to the problem. Some experts are urging government agencies to require your plant to install pollution control equipment, such as scrubbers, to reduce the emissions. The equipment will cost $100 million to $250 million and, in addition, will demand more energy. Such a regulation could force your plant to close at a time when it is struggling to compete with imported steel from Japan. You will urge the commission to continue to investigate the problem but to refrain from making any recommendations for regulating pollution emissions until it is proven without a doubt that your plant is responsible for acid rain.

Can the commission make valuable and sound recommendations even if some of the testimony was slanted?

Tanya/Carl Ryman, age 47, professor of architecture at MIT
As a professor of architecture, you feel strongly that something should be done to curtail acid rain if for no other reason than to spare some of the great buildings of the world from slow destruction by acid rain. In Athens, marble sculptures dating from the fifth century BC are turning to soft gypsum. The cause is air pollution—in particular acid-producing pollutants. The Parthenon, the Taj Mahal, and the Colosseum, as well as U.S. landmarks, including the Statue of Liberty, the Lincoln Memorial, and the Washington Monument, are being harmed by acid washings. The President's Council on Environmental Quality estimated in 1979 that the annual cost of architectural damage in the United States was nearly $2 billion.

Elliott/Brenda Jones, age 44, union representative of U.S. coal miners
Legislation has been introduced in the U.S. Congress to mandate a 10 million ton annual reduction in SO_2 by 1990 in 31 eastern states. Coal industry officials have testified before a Senate committee that 98 600 mining jobs would be lost in the Appalachian and Midwest coal regions if this legislation becomes law. Although you cannot vouch for the accuracy of these figures, you would like the commission to be aware of the effects this type of legislation would have on regional economies. You will ask the commission to weigh the costs and benefits carefully. The wrong legislation could destroy the already depressed economies of these regions.

mal pH by liming rather than by strict and costly pollution control laws. In liming, a neutralizing agent, such as lime or limestone, is added to a body of water. According to one estimate, liming 468 of the most acidic lakes in the Adirondacks to a pH of 6 would cost about $4 million per year. The addition of scrubbers to 50 of the oldest coal burning power plants east of the Mississippi is estimated to cost between $7 billion and $14 billion for capital costs and between $1 billion and $3.6 billion for annual operating costs. The benefits of liming appear within 1 year; scrubbers take 3 to 5 years to build. In Provincial's opinion, liming is an attractive option that the commission should consider strongly.

Maria/Gabriel Saracho-Mendez, age 29, representative of the U.S. Environmental Protection Agency (EPA)
The current EPA position is that some proposed methods of halting acid rain might foster worse problems. In addition, the agency contends that much acid rain is caused locally, contrary to the theory that acid rain in the northeastern United States and Canada is traceable to coal-powered plants in the Midwest. EPA believes acid rain requires additional scientific study before corrective measures are adopted. Some EPA researchers are studying how much acid rain is produced naturally.

Edward/Lynne Genito, age 54, president of Acme Steel Company, Pittsburgh
Your region of the United States, in particular its steel industry, emits high levels of SO_2. Many local weather

Helene/Marc Poisson, age 27, Canadian forestry official
Some researchers report that there is no proof that current levels of acid rain are significantly harming the terrestrial environment. In studies with simulated acid rain, the pH of the acid solutions used in the experiments that show adverse effects are usually lower than those normally occurring in rain. It is true that the leaching of nutrients from soil by acid rain could result in impoverished tree growth, but damage of this nature would take a long time to build up. Studies in Ontario do not yet show impaired tree growth. Although you are concerned about the problem of acid rain and you have no specific suggestions, you wanted to present recent research findings to the commission.

Penelope/George Laplace, age 45, president of Canadian Paper Products, Inc.
Most of the areas from which your company gets timber are located in regions where the pH of rain ranges from to 4.2 to 4.6, as identified by the Ontario Ministry of the Environment. Rain of this acidity is reported to reduce tree growth and to kill many plant species, including some of the pines that are the mainstay of your industry. Because the forest industry is Canada's largest, worth more than $10 billion annually in eastern Canada alone, you want the commission to recommend immediate restrictions on sources of SO_2 and NO_x, such as certain industries and coal-fired power plants. If acid rain is not reduced to

some extent, we can expect serious soil and forest effects in the next 25 to 100 years.

Kim/Sanford Johnson, age 42, freelance writer working on an article on acid rain
Although as a writer you try to remain objective, you have uncovered certain facts about acid rain that you think will interest the commission. The causes of acid rain include not only industrial pollution, but also automobile emissions, which account for much of the locally produced acid rain. Tall smokestacks built by utilities and industry reduce local pollution but transport acid droplets over great distances. Such information on dispersion should be useful in developing solutions to the problem.

Alice/Neil Shiring, age 35, representative for General Motors of Windsor, Ontario
The commission must consider all the facts before making decisions. Specifically, you will point out that in the United States, the President's Council on Environmental Quality stated that "insufficient knowledge exists about the explicit causes and total effects of acid rain . . . ." Any new and unnecessary pollution control measures required of industries such as yours could have devastating effects on the industry and eventually the national economy. You would be forced to pass the costs for pollution control on to the consumer. You remind the commission that the costs of pollution control are not only direct but also indirect because your suppliers pass on their added costs to you. With the increase in Japanese and European automobile production, North American industries can remain competitive only by keeping production costs down. You will point out to the commission that any acid rain legislation affects all parts of society.

Harold/Lisa Schmidt, age 32, public relations representative for Ohio River Valley Power Company
In response to an increasing demand for energy, your company plans to build a new power plant, either a conventional coal-fired or a nuclear power plant. Your company is interested in the commission's recommendations because they may affect which type of plant you build. Even if the public is strongly opposed, Ohio River Valley Power will decide to build the more economical nuclear plant if costly restrictions are imposed on emissions from coal-fired plants. Although your company is somewhat impartial on acid rain, you want the commission to consider your situation and to make recommendations as soon as possible.

Jeannette/Francois Durand, age 50, Deputy Minister of the Environment, Canada

---

Who should be responsible for the costs of reducing acid rain? Industry? Government? Citizens?

You plan to make the following statement on behalf of the Honorable John Roberts, Canadian Minister of the Environment:

"The solution to the acid rain problem is very straightforward. We must reduce drastically the amount of acid-causing pollution emitted in both countries. We already have the necessary technology: emission controls on smokestacks and cars and the ability to wash coal to cut down on the amount of sulfur released into the atmosphere. The argument against these solutions is cost, which admittedly is considerable; but the cost of not acting threatens forestry, tourism, and fishing. Building and automobile surfaces deteriorate each day. We may find that in the long run the cost to health has been great. No individual or group ever has the right to endanger the environmental health of a nation or a continent. Therefore, I urge you to recommend to both the Canadian and U.S. governments that immediate steps be taken to remedy the problem of acid rain."

Malcolm/Barbara Gordon, age 53, Governor of Kentucky
As a governor from the Ohio River Valley, you are well aware that acid rain is an international problem that should be addressed immediately. However, you will urge the commission to proceed thoughtfully and cautiously, examining all aspects of the problem before making recommendations. Your state is one of the leading coal producers in the United States. It has major sources of SO_2 and NO_x,

particularly in the northern industrialized area that borders Ohio. Restrictions on industries and electrical utilities could have serious economic repercussions in your state.

Marjorie/Michael Prince, age 30, Sierra Club lobbyist
The commission, you believe, must take immediate action against SO_2- and NO_x-producing industries. After all, research has shown that these industries are responsible for acid rain. Waiting for 5 or 10 years of government investigation may cost the environment dearly. In response to advocates of such remedies as liming, you will point out that liming requires 9 kg per surface hectare per year, which means several tons even for small lakes. Although liming may return the pH to normal, metal concentrations will still remain at levels toxic to fish. Also, the neutralizing effect wears off in 3 to 4 years. Liming may be a short-term remedy, but it is not a solution. Consider the logistics of dumping tons of lime into the thousands of lakes in Canada and portions of the United States!

Walter/Wendy Freeman, age 37, professor of ecology at Carlton University, Ottawa, Ontario
You would like to make the commission aware that acid rain threatens the environment and ultimately human health. Studies have shown a direct relationship between the severity of health problems and the level of air pollution as measured by the concentrations of suspended particulate matter, especially sulfates and SO_2 in industrial and urban settings. Acidification of water supplies could release metals from rocks, soils, or plumbing. In addition, the inhalation of sulfur particulates may cause chronic bronchitis and emphysema. Although none of these assertions has been proven, some evidence supports the validity of these hypotheses. ∎

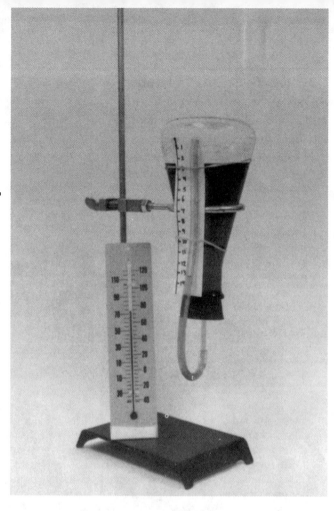

Air Apparent

Maybe you can't do anything but talk about the weather. Put a barometer in the classroom and that's all your students will want to talk about.

By David A. Harbster

I t's a sure thing. Bet any student who can tell you how a barometer works a round-trip, fun-filled week at Disney World. You can even raise the stakes. Pizza for life. You won't lose.

To most of us, I doubt that there's anything more familiar and less understood than a barometer. Everyone's heard of barometers, heard of barometric pressure, and heard about the instrument's measurements several times a day on every weather forecast. It means nothing to us. Unless

The top photograph shows a working model of the Cape Cod weather glass; the bottom, the Glendale barometer, invented by David Harbster.

we sail or live in a tornado path, few of us even care about barometers.

But we should. After all, a barometer forecasts fair and stormy weather. Some even believe it forecasts good and bad moods. Many of the teachers in my district attest to this. They say that the height of a barometer's column relates directly to their students' behavior, or, perhaps, misbehavior. But even if no real data support students' changing behavior with changing barometric pressure, studies show that bats and swallows are directly affected by low barometric readings. They avoid flying when it's low.

How a barometer helps bats get off the ground has to do with the air's pressure. A barometer is a precise gauge to measure air pressure. And measured air pressure comes from the gas molecules interacting with a solid or liquid surface.

Oddly enough, cold, moist air exerts less pressure than an equal volume of warm, moist air. There is a relationship between temperature and the moisture-holding capacity of air. The warmer the air, the more moisture it may contain. The moist air is less dense. As such, it exerts less pressure. And the barometer readings fall, giving low barometric readings for increased humidity and the possibility of rain. Dry air is more dense and therefore takes up more volume, creating higher pressure and higher barometric readings. Meteorologists use barometers to determine changes in air pressure. Barometers alone, however, cannot precisely predict the weather. Meteorologists need to know wind direction and temperature as well.

However, changes in air pressure can be observed in a classroom. Although the typical elementary classroom would not have a barometer as part of its equipment, it is easy enough to make one.

Stressing Air Pressure

I have made two barometric devices for the classroom that show air pressure changes. One of them is an adaptation of the classic liquid barometer, the Cape Cod weather glass. The other is what I call the Glendale barometer (named after my school district), which was the result of looking for a cheaper alternative to the Cape Cod version. Both of them are used in demonstrations with students from kindergarten through eighth grade.

To build the Cape Cod weather glass [see photograph on opposite page], you will need

- a ring stand,
- a three-in. diameter ring,
- a 500-ml. Erlenmeyer flask (or salad dressing bottle),
- a one-holed, number 08 rubber stopper,
- six cm. of five-mm. diameter glass tubing (or the glass part of an eye dropper),
- 30 cm. of flexible, clear plastic tubing,
- two rubber bands,
- water and food coloring,
- and a glass cutter or file to score the glass tubing for breaking.

Cut the length of tubing, and be careful when inserting the glass tube through the holed rubber stopper. I found

More Precise Conditions for Predicting the Weather

Wind Direction Pressure Reading	S to E falling	S to SW rising	SE to NE falling	SE to NE steady	
Weather Forecast	storm, gone in 24 hrs. 	soon clear, good days ahead 	rain for a couple of days 	fair for two days 	

	E to NE falling	S to SE falling	W rising	E to N low but rising	E to N low and falling fast
	rain in 24 hrs. 	wind, rain, in 18 hrs. 	clearing and colder 	cold wave 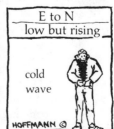	severe gale and rain 

HOFFMANN ©

that inserting the glass under running water works well. The water lubricates the glass and the rubber. Petroleum jelly and glycerin also work well. Never force the glass through. Gently twist the tube through the stopper. Wrap a towel around the tube to protect your hands.

Attach the plastic tubing—which can be found at any aquarium supply house—to the glass tube coming out of the flask. Fill half of the flask with colored water. Insert the stopper and invert the flask onto the ring stand. Bring the plastic tube up vertical to the flask. Any water that leaks out during extreme low pressure periods can be caught with a pie pan under the barometer.

The Glendale barometer costs about half of what the Cape Cod version does since it doesn't need the ring stand. A larger flask serves as its base and as a reservoir for the colored water. Also, it looks more "scientific," according to my students.

To make a Glendale barometer, you will need
- one 500-ml. Erlenmeyer flask,
- one 250-ml. Erlenmeyer flask,
- a 2-holed, number 06 rubber stopper,
- a 2-holed, number 08 rubber stopper,
- one 30-cm. piece of five-mm. diameter glass tubing,
- one 18-cm. piece of five-mm. diameter glass tubing,
- a three-cm. piece of five-mm. diameter glass tubing,
- water and two different food colorings,
- a gas burner,
- a glass cutter or file to score the glass tubing,
- safety goggles,
- and a ceramic pad or other suitable place to put hot glass tubing to cool down.

You can construct the Glendale in much the same way as the Cape Cod. The essential difference is in the use of a glass tube and some glass-tube bending. The glass bends are needed to form a column that projects out and upward away from the upper flask [photograph, page 12]. This minimizes the chances of it being bumped if it were too far from the barometer.

With the photograph of the Glendale barometer as a guide, make a glass plug with the three cm. of tubing by melting one end closed. Melt smooth the other edges to keep the students and yourself from being cut. This glazing should be done to any exposed glass tube ends. Then insert the plug into the smallest stopper.

Very carefully, bend the 30-cm. piece of glass to conform to the shape in the photo. Do not bend the glass directly in the burner flame. Just get it hot enough to be a little wobbly (turn it as you heat it), pull it away from the flame, and then bend it. Allow the glass to cool before inserting the rubber stopper. Insert both glass tubes into the stoppers.

Fill the larger, bottom flask completely with colored water. Fill half of the top flask with a different colored water. The glass tube and stopper then should be put into the smaller flask and inverted over and into the larger base flask. The water will very quickly rise into the side tube, and air space will be left in the upper flask. Make some minor adjustments of the lower stopper to get the water about halfway up the side tube. Attach a centimeter ruler or copy of a ruler to the tube. Use this to measure the changes in the column's height.

Students can now observe the effects of air's changing pressure as water rises or falls in the column. High pressure air will make the water drop, and low pressure air will make it rise. A warm classroom will cause the water column to rise three to five cm. You may want to attach a thermometer to the barometer to compare temperature changes with the water column's rise and fall. Be sure to keep the barometer out of direct sunlight to ensure accurate readings.

Remember, the actual reading on the column is not the most important data but whether the column is rising, falling, or steady. Stress to your students how crucial their observations are in determining general changes in atmospheric pressure. The class can even chart their observations and compare their data to the meteorological charts in the newspapers. Let them make broad weather predictions. They don't need a weatherman to know which way the wind blows. . . .

Resources
Bohren, Craig F. (1987). *Clouds in a glass of beer.* New York: John Wiley and Sons.

David A. Harbster, now a science education consultant and coordinator for the National Energy Foundation, was a science program specialist for the Glendale (AZ) elementary schools. Photographs courtesy of the author. Table, p. 13, by Thomas Hoffmann.

Phyllis Marcuccio

—————— Muhammad Hanif ——————

I s acid rain that is caused by factories in the Midwest destroying New England's forests? Is airborne pollution from the United States responsible for the death of Canadian lakes? And even if acid rain's role as an environmental hazard is firmly established, can our society afford to put an end to it? The controversy over acid rain is as heated as the issues are important. This makes it the stuff of headlines and editorials. It also makes acid rain an ideal vehicle for introducing students to the impact that scientific questions can have on society.

What Is Acid Rain?

The term *acid rain* is used to describe overly acidic precipitation of all kinds—sleet, frost, dew, mist, and fog—as well as dry particles that fall to the earth and form acids when they come in contact with moisture. The technical term for this phenomenon is *acid deposition*.

Scientists determine the acidity of precipitation by using the pH scale. On this scale, in which commonly found substances are classified from 0 to 14, 7 is neutral, readings above 7 indicate alkaline substances, and ones below are substances that are acidic. "Pure" rain water is lightly acidic, with a pH of 5.6 to 5.7, and any rain with a pH below that level falls in the category of acid deposition. Currently, the pH of the average rainfall in the northeastern United States ranges between 4.0 and 4.5.

But the numbers don't tell the whole story. Another factor to consider in determining the impact of acid deposition is the logarithmic nature of the pH scale, which means that there is a tenfold difference between one number and the next. A lake, therefore, with a pH of 6 is 10 times more acidic than one with a pH of 7; if the pH is 5, it is 100 times more acidic than a lake with a pH of 7; and if it is pH 4, it is 1,000 times more acidic than that same lake. Consequent-

ly, small dips on the pH scale indicate significant increases in acidity.

What Causes Acid Rain?

Natural events such as lightning and volcanic eruptions release substances into the atmosphere that can create acid rain, but the primary cause is human. Industries burning fossil fuels are an important source of the pollutants that lead to acid rain and so are emissions from automobiles and other vehicles. In the United States alone, these sources send 48 million metric tons of sulfur dioxide and nitrogen oxides into the atmosphere each year. These pollutants rise into high air masses, where they react with water vapor to form weak solutions of sulfuric and nitric acids. The acids then fall to Earth as acid precipitation.

Though scientists aren't certain about all the chemical reactions that cause acid rain, they do agree that the following reactions contribute:

• water combines directly with sulfur

Electric Power Research Institute

dioxide and forms sulfurous acid
$$H_2O + SO_2 \rightarrow H_2SO_3$$
• sulfurous acid combines with oxygen to form sulfuric acid
$$2H_2SO_3 + O_2 \rightarrow 2\,H_2SO_4$$
• water combines with nitrogen dioxide to form nitric and nitrous acids
$$H_2O + 2NO_2 \rightarrow HNO_3 + HNO_2$$

Possible Effects of Acid Rain

Some scientists believe that acid rain is changing the level of acidity in certain freshwater lakes and thereby adversely affecting aquatic life. When the pH of a lake is below 5.6, fish suffer from a lack of calcium that weakens the skeleton; thus, acidic lakes often contain many humpbacked, dwarfed, or

Muhammad Hanif is an associate professor in the Department of Early and Middle Childhood Education at the University of Louisville, Kentucky.

otherwise deformed fish. Acidity also releases aluminum from the soil that can then enter the water and ultimately the gills of fish. Though it may take years, the aluminum can build up in the gills until a fish eventually suffocates. Evidence also suggests that high acidity impairs reproduction and decreases the viability of eggs and young fish.

Acidity also affects aquatic plant life. It destroys microorganisms; consequently, organic matter decomposes more slowly, decreasing the number of plankton and thus disturbing a vital link in the food chain. Lakes that seem to be particularly susceptible to acid rain are those that do not have buffering substances, such as limestone or dolomite, to neutralize acidity.

Ironically, spring, the season of birth, may now be a season of death for some lakes because of acid rain. When the snow melts, it carries high concentrations of acid into the lakes, causing an "acid shock" at a time when much aquatic life is young and vulnerable.

Scientists still haven't determined just how extensive the damage to aquatic life has been as a result of acid rain. Some argue that there are too many conflicting and questionable studies of acid rain to draw definite conclusions about its effects; others do not accept the current evidence because they think there is too little long-term data; and some suggest that the damage may be a result of other forms of pollution.

Scientists are also debating the effect of acid rain on terrestrial plants. Some plants, like sugarcane or corn, can react favorably to acidity; others show no reaction at all. This is especially true of plants in soil that is well buffered. But some soil, such as that in the northeastern United States and Canada, has few buffers; and many scientists believe that acidity in these areas can damage plants. When the acids release metals such as aluminum and magnesium from the soil particles, the metals can enter the roots of the plants and inhibit their growth. (These metals can also contaminate water supplies and fish so that they are unfit for human consumption.)

Other theories state that acid rain destroys the protective coating on leaves, making them susceptible to infection and insect infestation and that excess acid deprives the soil of essential nutrients, such as calcium, potassium, and sodium. Also, some evidence suggests that acid retards decomposition and kills earthworms, both of which influence soil fertility. Again, the extent of possible damage remains uncertain.

According to some scientists, acid rain may be one of the major causes for rapid corrosion of structures such as the United States Capitol Building and the Statue of Liberty. However, this, too, is a debated issue. Though it's well established that acid does corrode many building materials, some scientists believe that other forms of pollution may be equally or more responsible.

Preventing Problems

There are some ways to limit the possible effects of acid rain. For example, the Swedes have spread lime on and around lakes that have become overly acidic to neutralize the effects of acid deposition. Selective breeding of fish resistant to acidified water is also possible: the Canadian brook trout, for example, is especially tolerant. In cities, protective coatings can prevent the corrosion of buildings and monuments that may be caused by acid precipitation.

Another solution is to eliminate, or at least reduce, the release of substances that create acid rain. Some power plants that burn fossil fuels now put scrubbers in their smokestacks to remove the sulfurous materials from the residue. Scrubbers spray the effluent traveling through a smokestack with water containing chemicals, such as lime, that will react with sulfur dioxide and convert it into by-products that are easily removable. The plant can then clean out these by-products and dispose of them or use them as one of the ingredients needed in road surfacing.

But these preventive measures are costly, and some think that we know too little about acid rain to take action.

Getting Students Involved

The key to resolving the debate on acid rain is public involvement. Groups interested in confronting the issue are active at the local, state, national, and international levels. Your students, too, can get involved. Here are some hands-on activities that will give them a better understanding of the issues.

1. *Finding the pH of different solutions.* Explain the pH scale to your students, and have them use litmus paper to determine the pH of various substances, such as those listed below. Their results should be close to those that follow:

lemon juice 2.0
vinegar 2.2
apple juice 3.0
milk 6.6
distilled water 7.0
baking soda 8.2
milk of magnesia 10.5
ammonia 11.5

After making their measurements, students may want to put their results on a graph of the pH scale. (See Figure 1.)

2. *Finding the pH level of rainwater and freshwater.* Have students collect samples of rainwater from different neighborhoods as well as water from nearby lakes or streams, using thoroughly cleaned bottles or cans. They should test the pH of each sample immediately and then label the sample with the pH reading, date, time, location, and water source. Have them make their collections for several months or throughout the school year, recording the data on graphs so they can easily note any differences or trends. (See Figure 2.) Once they have completed the collections, have them analyze the data. Some of the questions they should answer are

- What trends do you notice when you compare the acidity level of rainwater or freshwater collected from the different places?
- In which month is the pH level lowest for rainwater? For freshwater?

Canadian Embassy

(Above) Sulfur oxides from power plants in the Ohio Valley may damage forests and lakes as far away as northern Ontario, northern Quebec, and the Northwest Territories of Canada. Shallow soils, rock outcroppings, and granitic bedrock—together with heavy evergreen cover—make these regions extremely susceptible to damage by acid rain.

- In which month do you find the highest pH for each?
- Does time of day affect the pH of either rainwater or freshwater collected at a specific spot?
- Does distance from an industrial plant affect the acidity of your samples?

3. *Investigating the effects of acidity on plants.* Have students fill four plastic trays with vermiculite (a hydrous silicate); label them A, B, C, and D; and plant each one with an equal number of lima bean seeds. Students should let seeds sprout and grow into plants for about four weeks, watering them with solutions

of different acidity. For tray A have them use regular tap water; for tray B, a solution of 1 part water and 1 part vinegar; for tray C, a solution of 5 parts water to 1 part vinegar; and for tray D, a solution of 50 to 1. Have the students observe and record the growth rate of the plants in each tray and explain any differences they notice.

4. *Determining the effects of acidity on foliage and roots.* Use two equally healthy potted begonias for this experiment. Have the students mist one plant daily for approximately one month with tap water and the other with a water-vinegar solution of pH 4. Water each plant with tap water as necessary. The students can observe and record the differences in the growth of the two plants and the appearance of their leaves. What conclusions can they draw from this experiment?

In another comparison involving the effects of acidified solutions on plants, place a cutting from a coleus, marigold, or begonia plant in tap water, and another cutting in a solution of vinegar and water with a pH of 4. Observe the effects on roots over a period of three weeks.

5. *Illustrating and discussing the causes and possible effects of acid rain.* Using their own artwork or articles and pictures from newspapers and magazines, students can make posters or bulletin board displays that illustrate theories about the causes and effects of acid rain. Ask students to present their posters and displays to the class and discuss their reasons for showing the issue as they do.

If interest is high, organize a debate or help students stage a "public forum" in which they play the parts of people on various sides of the acid rain controversy. For instance, one student could be an environmentalist, another a New England farmer, and a third a member of Congress from an industrial town in the Midwest. (See Rodger Bybee, Mark Hibbs, and Eric Johnson's "The Acid Rain Debate" in the April 1984 issue of *The Science Teacher*, pages 50–55, for suggestions on setting up this kind of forum.)

A unit on acid rain presents a teacher with the advantages and disadvantages of a controversial and timely subject. It

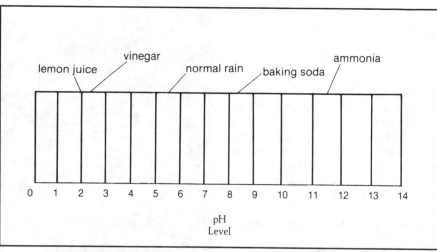

Figure 1. **Some Common Substances Located on a pH Scale.**

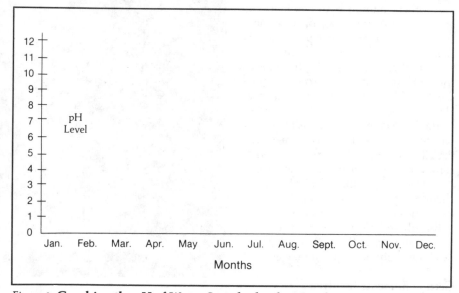

Figure 2. **Graphing the pH of Water Samples by the Month**

is probably easier, and it is certainly safer, to stick with subjects on which there is general agreement. But getting students to look at a subject on which people don't agree is a sure way of showing them that science is both important and interesting. And it will be a valuable lesson about the kinds of problems students will have to deal with as citizens and voters.

Resources

"Acid Rain Research: A Special Report." *EPRI Journal* 8, November 1983.

"Acid Rain." *Arizona Energy Education* 6, December 1983.

"An Introduction to Acid Rain." *Science Scope* 7:6–9, February 1984.

Bybee, Rodger. "Acid Rain: What Is the Forecast?" *The Science Teacher* 51:36–40, 45–47, March 1984.

Environment Canada. "Downwind: The Acid Rain Story." Ottawa: Minister of Supply and Services Canada, 1982.

Gibbons, D. L. "Acidic Confusion Reigns." *SciQuest* 55:10–15, January 18, 1982.

Jacobs, Madeleine. "Taking the Measure of Acid Rain." *Smithsonian Institute Research Reports* 39:1, 8, Spring 1983.

Kerr, R. A. "Pollution of the Arctic Atmosphere Confirmed." *Science* 212:1013–14, May 1981.

LaBastille, A. "Acid Rain: How Great a Menace?" *National Geographic* 160:652–81, November 1981.

Nader, R. et al. *Who's Poisoning America?* San Francisco: Sierra Club Books, 1981.

National Wildlife Federation. "Acid Rain: A Teacher's Guide." Washington, 1983. (Available from the Federation, 1412 16th St., N.W., Washington, DC 20036 for $1. Ask for item 79678.)

Ohanian Susan. "Will April Showers Kill the Flowers?" *Learning* 12:80–88, April/May 1984. (Includes an extensive bibliography.)

Introducing Isolines

A weather-wise activity.

by Michelle Akridge, Barbara Lary, and Gerald H. Krockover

Every day, people come into contact with all types of maps. You probably saw at least one map today, perhaps the weather map on the morning news. Many maps use special features called isolines to convey information. Isolines show where certain features—such as altitude—are constant over the area displayed on the map. Several of the most common types of isolines are isobars, which show equal pressure, and isotherms, which show equal temperature. (See the map on page 33.)

Many students have difficulty understanding how isolines are determined. Most textbooks present the concept of isolines without explaining how isolines are derived. The following activity will provide your students with a concrete example of how isolines translate into the maps that they see everyday. In the first part of the activity, students will construct isolines with string around an array of push pins placed in a cork board. (See Figure 1.) This exercise will give them the knowledge about isolines which they need to analyze real weather maps in the second part of the activity.

The isoline activity will require some setup on your part. First, you will need to gather the following materials: three cork boards, push pins (in five different colors), colored pencils (preferably the same colors as the pins),

Michelle Akridge is an atmospheric science education specialist in the department of earth and atmospheric sciences and Gerald H. Krockover is a professor in the departments of earth and atmospheric sciences and education at Purdue University, West Lafayette, IN 47907. Barbara Lary is an earth science teacher at North Central High School, 1801 E. 86th St., Indianapolis, IN 46240.

graph paper, three balls of string, scissors, and a variety of maps from the local National Weather Service Office and/or from newspapers.

Illustrating isotherms

To prepare the isoline arrays for the first part of the activity, place colored push pins on each of the cork boards following the patterns shown in Figure 1. These patterns represent three basic isoline patterns commonly found on weather maps.

Each of the boards will serve as a work station during the activity. Divide the class into three groups. Have the groups rotate so that each one visits all three stations.

At each of the stations, students connect push pins of the same color with string. For example, at work station two, students will connect all the yellow pins on the board. Have them start by loosely knotting the string to a yellow pin close to the edge of the board. They should then connect the string to the nearest yellow pin by looping the string around that pin once. The students should continue looping the string around the closest "unlooped" pin, until all the yellow pins are connected. Warn them not to pull the pins out of the board while working.

When the students have finished connecting all the pins of one color, have them cut off the ends of the string. They should repeat this procedure for the other colors on the board.

Students should follow certain rules when connecting the push pins with the string. First, they should make sure that the strings do not cross themselves or each other. Second, they should see that the lines made by the strings are similar to the curves.

When students have finished connecting the push pins with the string, they should copy the pin points and lines as closely as possible onto a piece of graph paper, making sure to note the colors of the pins. All lines should be labeled with the appropriate color, or drawn with the appropriately colored pencil. Your students should notice that the lines form specific patterns on the paper. Ask them what shapes the lines make. The students probably will respond that the lines look like either waves or rough circles. Before the students move on to the next work station, ask them to remove the strings at their current work station.

Once the students have visited all three work stations and constructed their isolines, it's time to introduce them to the idea that similar lines, when drawn on a weather map, represent a range of locations with constant temperature or pressure. Tell them that each work station represents one weather pattern and that each color corresponds to a certain temperature or pressure value.

Work Station 1 shows a pattern of temperature. (See Figure 1.) The push pins on the board represent weather stations and the colors of the pins correspond to the temperatures at the stations. The lines are isotherms—lines of constant temperature. Students should plot the following temperatures at the corresponding points on their map: yellow, 22°C; green, 20°C; and blue, 18°C. The students will be interested to know that a real temperature map might look similar to the one they have made.

The pattern at Work Station 2 represents atmospheric pressure, which is measured in millibars. Again, the pins represent weather stations, and

—Art by Max-Karl Winker

National Science Teachers Association

the colors correspond to the atmospheric pressure at these stations. The lines are isobars—lines of constant pressure. Students should plot the following pressure values at the corresponding points on their maps: yellow, 1012 mb; green, 1010 mb; and blue, 1008 mb.

The pattern at Work Station 3 also represents atmospheric pressure, so the lines are again isobars. Give students the following pressure values to plot: blue, 996 mb; green, 1000 mb; yellow, 1004 mb; orange, 1008 mb; and red, 1012 mb.

When the students are done with their mapping, ask them the following questions.

• What is the temperature difference in °C between the lines that you have drawn for Work Station 1? What is the pressure difference in millibars between the lines that you have drawn for Work Stations 2 and 3? (The difference at Work Station 1 is 2°C, the difference at Work Station 2 is 4 mb, and the difference at Work Station 3 is 4 mb.)

• How would you describe the pattern that you drew at each work station?

• Is the atmospheric pressure high or low in the center of the map for Work Station 2? Why? (High, because the pressure values increase as you move toward the center.)

In the second part of the activity, students examine the real weather maps that you collected. Have the students read the map legends, dates, and time, and determine what weather conditions are being displayed on the map. Then ask them to answer the following questions. Either set up a class discussion, or have the students work in smaller groups and record their answers.

• What are the sources for the weather maps? What other sources do you know about?

• What information is represented on each weather map?

• What are the differences between the maps?

• What is your location on each of the maps? (It may not be marked.) What are the weather conditions reported or predicted for the day the map shows?

• How are the isolines useful? Compare a board and pins with no string (lines) to one of these maps.

• How do you think these maps

were prepared? Relate these maps to your isoline activity.

You can develop extensions of the isoline project that allow students to deal with the problem of drawing isolines with actual numbers, so that the activity is no longer a matter of "connecting the dots." One good source of data for constructing an isotherm map is the *USA Today* "Weather Across the USA" page, which lists temperatures from cities all around the country. Students can take the temperature values from the page and place them on the appropriate spots on a blank map. After drawing isotherms on this map, students can check their results

by comparing them to *USA Today*'s map of the country. The newspaper uses color shading to designate areas that fall within certain temperature ranges. The isotherms are the boundaries between the different shades.

The isoline activity can serve as a beginning in mapping and analyzing weather data. It also can serve as a springboard to understanding other maps that use isolines, such as topographic maps, for example, which use contour lines. Our experience with this activity leads us to believe that it can make a concept that is often difficult for students to grasp more understandable and enjoyable. ∎

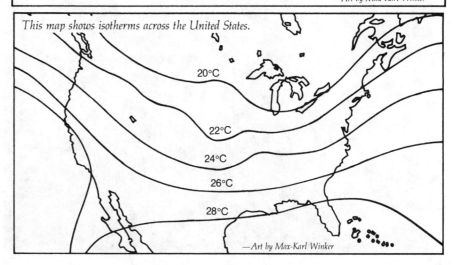

Figure 1. Cork board and push pin setups.

Follow these patterns when setting up the push pin arrays for isoline construction. The letters represent the colors of the push pins (B, blue; G, green; O, orange; R, red; Y, yellow).

Work station one

Work station two

Work station three

—Art by Max-Karl Winker

This map shows isotherms across the United States.

20°C

22°C

24°C

26°C

28°C

—Art by Max-Karl Winker

Balloons and Blow Bottles

David A. Harbster

One of the things that is difficult about discussing air is that it is both omnipresent and mysterious. We breathe air, and everyone has heard the term *air pressure*. But what does the evidence of our senses tell us about air? Students have felt the weight of water, and they probably know that if they were to stand at the bottom of the ocean, the pressure exerted by the weight of the water would crush them. We live at or near the bottom of a sea of air. Do we feel the effects of its pressure? Does air really take up space? How can we talk about any of these things when we can't even see what we're talking about?

A good place to begin answering some of these questions is by looking at phenomena that students have undoubtedly noticed and accepted without giving them any thought.

The Popping of the Cork

For instance, when we open a carbonated beverage—soda pop or beer—there is a hiss and sometimes even a spray of liquid. And with champagne there is the well-known pop and gush. This display occurs because the beverage has been canned or bottled with its contents under pressure and is therefore at higher pressure than the surrounding air. When the container is opened, there is a spontaneous equalization of pressure.

David A. Harbster is a science program specialist with the Glendale (Arizona) Elementary School District. Photographs courtesy of the author. Diagrams by Hana Ponds.

See if students can tell you what would happen if the contents of a pop bottle were under the same pressure as the surrounding air. Would there still be a hiss? (See box.) And what if the air pressure surrounding the bottle was greater than the air pressure in the bottle? In which direction would the air rush? Introduce students to the word *implode*. Why would there be a danger of implosion in this situation?

And the Popping of the Ears

Another familiar example of air pressure in action is the popping of the ears that occurs when people flying in an airplane make a quick and drastic change in altitude. When this change occurs, a difference in pressure is created between the air in the middle ear and the air outside the body. This difference is equalized either by air rushing into the eustachian tube or escaping from it by way

the can quickly by immersing it in a pan of cold water. As soon as the contents of the can have cooled, the can will collapse. What has happened? (The air has already been driven out by the steam. When the steam condenses as water, it leaves a partial vacuum, and the can collapses as the outside air pressure pushes in from all sides.)

A second activity, which involves inflating a balloon inside a bottle, is less spectacular than the collapsing can, but it lends itself to variations that are very useful in helping students understand some of the properties of air—especially if you use a device that I call a *blow bottle*.

Now Blow—Hard!

To construct a blow bottle, you will need a serrated knife, a cutting board, some silicone glue, a plastic milk or spring-water jug, and a one- or two-liter plastic beverage bottle with a screw-

We live at the bottom of a sea of air. Do we feel the effects of its pressure? Does air really take up space?

of a valvelike flap, and causing the eardrum to bend inward (if the plane is descending) or outward (if the plane is ascending). People chew gum and mothers make sure that their babies have something to suck on because this kind of activity keeps the eustachian tube open and the air pressure continually equalized.

The Collapsing Can

Two classic demonstrations will give students a chance to see some properties of air for themselves. The first illustrates, rather spectacularly, the principle involved in the imploding pop bottle mentioned earlier.

Wash a can with a screw top (like a maple syrup can), and heat a small amount of water in the can. (Caution: Don't use any can that has contained flammable material—paint or lighter fluid cans, for example.) As soon as the water boils and steam billows out, cap the can tightly, using gloves or a pot holder to protect your hands. Then cool

on cap. (Avoid jugs that have shiny or slick surfaces because silicone glue may not adhere to them very well.) Remove the label from the jug by soaking it in hot water. Cut off the top of the plastic beverage bottle directly below the flange. (See Figure 1.) Cut a hole of the same diameter as the mouth of the beverage bottle in the side or bottom of the jug (or in another beverage bottle). Spread silicone glue in a ring about 0.5 centimeters (cm) wide on the outside perimeter of the hole you have just made. Place the stem of the cut-off top in the ring of glue, and rotate it gently to spread the glue evenly. Then to ensure a secure seal, add another ring of glue 0.4–0.8 cm wide between the neck of the bottle and the jug. (See Figure 2.) Allow the bottle to dry undisturbed in a warm place for at least 24 hours.

Begin the demonstration of air pressure by screwing the cap securely on the bottle top. Then insert a sturdy balloon into the top of the jug, arrange the end of the balloon over the mouth

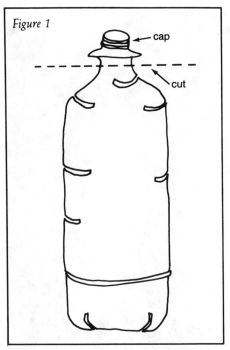

Figure 1

cap

cut

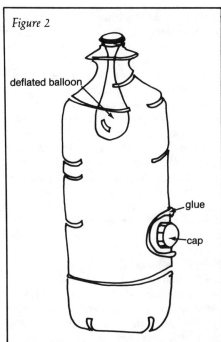

Figure 2

deflated balloon

glue

cap

person preparing to suck lowers both tongue and jaw, which enlarges the volume of the inside of the mouth. This, of course, lowers the air pressure inside the mouth. In this case, the air from outside exerts its pressure on the liquid in which the straw is placed and forces the liquid into the drinker's mouth.) What would it be like to suck on a straw if you were in a setting where the pressure was very low—like on Mount Everest—or very high—like in a diving bell at the bottom of a river? See if your students can figure that one out.

Resources
Goldstein-Jackson, Kevin. *Experiments with Everyday Objects.* Englewood Cliffs, N.J.: Prentice-Hall, 1978.
Grosser, Arthur E. *The Cookbook Decoder, or Culinary Alchemy Explained.* New York: Warner Books, 1981.
Hewitt, Paul G. *Conceptual Physics.* Boston: Little, Brown, 1977.

of the jug, and blow into the balloon. Why does the balloon inflate a bit and then stop? If students think it's because there's something wrong with the balloon, remove the balloon from the jug and inflate it. Someone may be able to come up with the idea that *the air* in the bottle is taking up space, and as it is compressed by the balloon, it, in turn, exercises more and more air pressure. But if that answer is not forthcoming, unscrew the bottle cap and blow again. Now the balloon will inflate. What has happened? The air pressure in the bottle, which was increased by inflating the balloon, has decreased with the escape of air and has now become equal to the air pressure outside the bottle. (Remind students of the example of the eustachian tube.)

When students seem to understand how air pressure has worked in these two cases, tell them that you're going to ask them to predict what will happen to the balloon in some other situations:

• What will happen if, after inflating a balloon in the blow bottle and then screwing on the cap, you remove your mouth from the balloon? (As the balloon begins to deflate, the air decompresses, and the air pressure inside the bottle becomes slightly lower than the air outside. As a result, the outside air,

which is under higher pressure, tends to move into the balloon, and this keeps the balloon partially inflated.)

• What will happen if you attach a second balloon firmly to the opening in the side or bottom of the blow bottle with rubber bands and then blow into the first balloon? (The first balloon will inflate, and then, because of the air pressure set up inside the bottle, so will the second balloon.)

The Last Straw

Since blow bottles are inexpensive and easy to make, individual students, or pairs of students, can construct their own and replay these demonstrations, as well as variations that you or they devise. While they experiment, encourage them to think about how the principles illustrated here apply elsewhere in their daily lives. For instance, the blow bottle gives an insight into how the lungs work. (As the ribs expand and the diaphragm, which is shaped like a dome, moves downward, the air pressure in the chest is decreased and air from outside rushes in to equalize the pressure by inflating the lungs.)

And, to return to the soda pop with which we started, an understanding of air pressure can also explain how people suck up their drinks with a straw. (A

What? No Fizz?

A legendary story of underwater festivities illustrates what happens when people drink under pressure.

A group of dignitaries met to celebrate the joining of two shafts of a tunnel under a river. (If one account is correct, the river was the Thames and the year 1827.) The champagne seemed to be flat when it was opened, so the dignitaries toasted the occasion in champagne without bubbles.

But when they returned to the surface after the celebration was over, the champagne—now in their stomachs—became very bubbly indeed, and one man had to hurry back down to the tunnel, to undergo "champagne recompression."

The reason? Carbonated beverages, which are bottled under pressure, contain gasses dissolved in the liquid. When their bottles are opened, the gasses bubble out of the solution, causing the familiar fizz. But when a bottle is opened under circumstances where the air pressure is comparable to the pressure under which the beverage was bottled, the gasses remain in solution and the beverage is flat. Flat, that is, until the drinkers return to normal air pressure.

Why Is the Sky Blue?

————— **William G. Lamb** —————

When middle school students find out that I'm interested in astronomy, they often ask me to explain two common phenomena: why the sky is blue and why the Sun seems squashed on the horizon right before it sets. I can easily tell them that the sky is blue because the fine particles in the atmosphere scatter more blue light than red. I can just as easily explain that the Sun appears squashed because it is already below the horizon and its image is being refracted by the atmosphere. But both explanations are just so many words to the students. Some of them may be able to parrot what I've said, but very few actually understand the explanations. Consequently, I have searched for demonstrations to illustrate the principles involved, and here are two good ones.

Why Is the Sky Blue?

To answer this question, you will need to demonstrate a principle called the Tyndall effect. First, get yourself an aquarium tank or similar container, preferably one with a clear bottom, although an opaque-bottomed one can be used). Ideal dimensions for the tank are 150 millimeters (mm) high, 150 mm wide, and 250 mm long. If you cannot find a tank, you can make one from pieces of sheet glass glued together with the silicone rubber that is available at most shops that sell tropical fish. Once you have a suitable tank, try one or both of the following demonstrations.

Using water with milk. Fill the tank with water and adjust the beam of light from a projector so that the light passes through the water to form a sharply focused spot on a screen or white piece of paper attached to the wall. (See Figure 1.) Add milk, a few drops at a time, until the beam of light is clearly visible through the length of the tank. Continue to add milk a little at a time, and observe the color of the water from the side of the tank and the color of the spot on the screen. As more milk is added, the light scattering out through the side of the tank appears bluish while the spot on the screen becomes progressively redder.

Figure 1

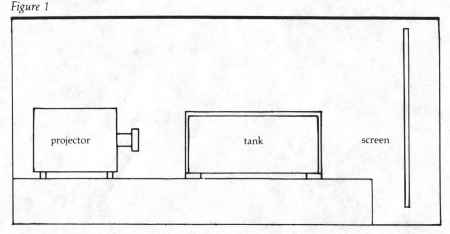

Figure 2

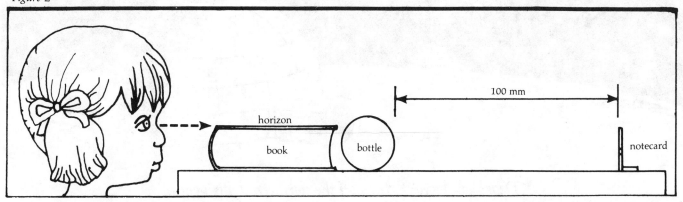

Earth at Hand

Using sodium thiosulfate and hydrochloric acid. Prepare 5 liters (L) of sodium thiosulfate solution (known as hypo to photographers) by dissolving 100 grams in 5 L of water. Put the solution in the tank and focus the projector light through it onto the screen, as before. Add five to ten drops dilute of hydrochloric acid and stir. (Wear rubber gloves and a lab apron when working with hydrochloric acid.) A colloidal suspension of elemental sulfur slowly precipitates out, and the light being scattered out the side of the tank appears bluer and bluer as the spot on the screen becomes yellow, orange, and, finally, red.

When preparing the "blue sky" demonstration, darken the room as much as possible. You can use a slide projector or an overhead projector as your light source, but an aquarium with an opaque bottom will not work well with an overhead projector. If you use a slide projector, you may want to make a special slide by mounting a piece of aluminum foil in an empty slide mount and

Be prepared the next time a student asks why the Sun looks squashed at sunset or why the sky is blue.

then carefully punching a hole about 18 mm in diameter through the center of the foil slide. This produces a smaller light beam and a more pronounced effect.

Why Is the Sun Squashed?

To demonstrate the refraction of the setting Sun, you will need a clean, colorless, transparent bottle; a cork to fit it; a notecard with a circle 10 mm in diameter drawn in the middle of it with a red felt tip pen; and some books. Fill the bottle with clean water and cork it

tightly. Arrange the materials as shown in Figure 2. A student looking along the surface of the book, which represents the horizon, will then be able to see the red spot on the card even though the spot is below the plane represented by the top of the book. Now remove the bottle. What happens?

Refraction can be a difficult concept to explain, and this demonstration is a simple and dramatic method of introducing it as well as a way of explaining the "squashed Sun" phenomenon.

Once you have accumulated the materials needed to perform these two demonstrations, you can repeat them on a moment's notice. So be prepared the next time a student asks you why the Sun looks squashed at sunset or why the sky is blue.

William G. Lamb is science head at Oregon Episcopal School, Portland, where he also teaches chemistry and physics. Artwork by Darshan Bigelson.

"Dear, you forgot to turn off the northern lights again."

Airborne Particles

The humble vacuum cleaner makes a first-rate air sampler in this environmental science activity.

by Carl F. Ojala and Eric J. Ojala

They leave grit on our windowsills, reduce our driving visibility, irritate our allergies, and contribute to acid rain. They are airborne particulates, minute particles in the air, existing in tremendous numbers, falling into a variety of types, and occurring both naturally and from industrial processes.

Particulates range in size from debris from tornadoes and volcanoes to pollen grains, salt, dust, and sand blown up by the wind to smoke and soot from fires to submicroscopic materials such as silver or lead iodide and fine particles of sea salt. An investigation of airborne particulates can help science students understand more about the environment. By determining the size of particulates, as well as their relative concentration in the air, you can generally gauge the air quality of a region.

But the method of investigation is as important as the results. One procedure calls for coating a microscope slide with a sticky substance, clamping the slide to the end of a stick attached to the top of a car, and driving around to collect samples [2]. To devise another way of collecting the particulates, one that doesn't require a car, and then to measure their concentration under different atmospheric conditions requires students to exercise their ingenuity and their knowledge of science. Here's a solution that uses a vacuum cleaner.

The first step is to determine the amount of air that passes through the vacuum cleaner in 1 minute. By attaching a large plastic trash bag of known volume to the exhaust vent and timing with a stopwatch, the students can establish how long it takes to fill the bag with air. Then, using a series of conversion factors, they obtain the value they seek, in cubic meters. Comparing this volume with the volume of air we breathe each day (about 7 cubic meters) will give the students a clearer picture of particulate concentration in the environment.

Next, the students move outdoors to collect the particulates. They coat a microscope slide with an adhesive such as Mayer's albumin or petroleum jelly. Then they open the vacuum, and anchor the slide by making a pocket with duct tape at a convenient spot anywhere in the air stream, for example, at the opening where the hose attaches. When the vacuum is closed and turned on, air is sucked in through the intake vent, over the slide, and out through the exhaust vent. After the slide is exposed to the air, the students open the vacuum again, remove the slide, coat it with permanent slide mounting medium to keep the particulates in place, and add a cover slide.

Once they perfect their procedure, the students collect samples of air under a variety of atmospheric conditions. Weather conditions are most changeable during the fall and spring months. A month or two should be sufficient for collecting samples.

After collection is complete, they take several pictures of the slides through a microscope. As you can see from Figures 1 and 2, many particulates are round, but others have irregular shapes. Some, such as fine smoke particles from burning, are invisible to the unaided eye; others like fly ash or dust from industrial sources are large enough to restrict visibility. Airborne particulates very in size from about 0.001 micrometers to about 100 micrometers. Still, those at the high end of this range are only slightly wider than a human hair.

Particulate statistics

The largest particulates, which are the least in number, form from the fragmentation of still larger particles by mechanical breakdown, weathering, abrasion, or other types of attrition, finally becoming small enough to be suspended in the air. Examples are mineral dust from industrial grinding and ashes from incinerators and other fires.

The smallest particulates, which are greatest in number, form from the condensation and crystallization of hot gases from cars, steel plants, fires, or volcanoes, yielding very small ions and molecular clusters [2].

After developing the film, the students count the number of particulates visible in the photograph. It takes a series of careful calculations to determine how many particulates were present in the air at the time of each sample. First, the students calculate the area of the slide, then they divide that value by the area visible in the photographs. Let's say the area of the slide was 1000 mm² and the area

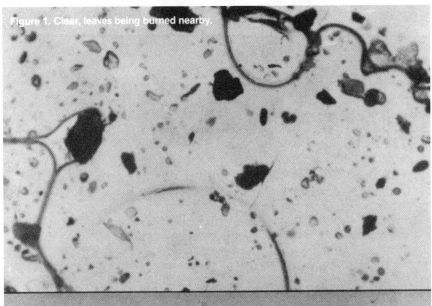

Figure 1. Clear, leaves being burned nearby.

Figure 2. Overcast and rainy.

—Photos by the author

shown in the photograph averaged 5 mm². This ratio would be 200. By multiplying the number of particulates counted in the photographs by 200, the students determine the number of particulates present on the entire slide. For example, if the particulates averaged 7 in the photographs, that would mean that 1400 were present on the whole slide.

Next, they find the area of the opening through which the air was drawn not covered by the slide. Let's say the area of the opening not covered by the slide was 450 mm². Simple addition and division determines that the slide covered 69 percent of the open-

Carl F. Ojala is a professor of geography at Eastern Michigan University, Ypsilanti, MI 48197. Eric J. Ojala is a student at Brighton High School, Brighton, MI 48116.

ing. So, to find the number of particles sucked in by the vacuum during the collection period, the students multiply 1400, the number collected on the slide, by 31 percent. Then they add the result, 434, to the number collected on the slide, 1400, for a total of 1834.

After finding the number of particulates collected in each sample, the students graph the data against different atmospheric conditions. Among the variable atmospheric conditions that can be measured are wind direction, wind velocity, air pressure, temperature, relative humidity, cloud cover, and precipitation. Students should be reminded that all these atmospheric conditions need to be recorded at the time the sample is taken.

This is a good time to discuss how researchers analyze data using statistics. For example, the students could run a regression analysis program on a computer with the data from each graph and then add the regression lines to their graphs to indicate the average position of all the data shown. A regression line can be used as a prediction line; that is, if a value on the x-axis is known, then a corresponding value can be determined from the y-axis.

The weather factor

At the close of the investigation, students can draw some conclusions about the effect of weather conditions on the number of particulates in the air, or air quality (as long as they understand that researchers can make absolute statements only under strict experimental conditions and after gathering and analyzing voluminous data). Generally, on clear or windy days, the number of particulates in the air will be relatively greater than on still or damp days.

This makes sense, as the air is continually being cleansed of particulates by a variety of processes, primarily processes associated with the weather. During inclement weather some of the relatively larger particles become cloud condensation nuclei. As soon as a cloud droplet forms, it serves as a gathering place for the smaller particles present. Often thousands of fine particulates will encounter and stick to one cloud droplet [2]. Some particles are captured, or swept up, by raindrops or snowflakes as they fall through the air. In fact, precipitation accounts for 80 to 90 percent of particulates removed from the air [1].

Although this study describes a methodology for collecting and measuring particulates, it could be extended to investigate issues more complex than air quality, such as acid rain and air pollution. Determining the identity of the materials collected on the slides would be another direction to explore. But sparking students' curiosity about earth science is the first step, and collecting particulates with a vacuum should do the trick.　■

References
1. Gedzelman, S.D. *The Science and Wonders of the Atmosphere.* New York: John Wiley and Sons, 1980.
2. Schaefer, V.J., and J.A. Day. *A Field Guide to the Atmosphere.* Boston: Houghton Mifflin Co., 1981.

A Crushing Experience

George Kauffman

Here is an old standard plus a new variation. Both are dramatic demonstrations of the (atmospheric) pressure under which we live.

Air Pressure—A Simple Demonstration

The pressure of the atmosphere (*ca.* 14.7 psi, 760 torr [mm Hg], or 1.0×10^5 Pa for SI buffs) can be demonstrated in a simple, striking manner with no specialized apparatus as follows: Add water, several centimeters deep, to a one-gallon, flat-sided metal varnish can (right parallelopiped) or a mimeograph duplicating fluid can. Place the can on a tripod over a Bunsen-burner flame, and boil the water vigorously until steam escapes profusely (see Figure 1*a*). A hot plate may be used, but it requires more time. Turn the flame off, and *immediately* and *quickly* replace the screw cap, using an asbestos or heat-resistant glove. A rubber stopper may be used instead of the original screw cap. (Ed. note: A No. 5 rubber stopper works fine for duplicating-fluid cans and alleviates the problem of a leaky gasket in the original screw cap. Also, I recommend a rubber stopper for safety reasons in case you get excited during the course of the demonstration and forget to turn off the flame.)

Since the air within the can has been expelled by the steam, a partial vacuum is created as the steam condenses. The pressure of the outside air, which is no longer counterbalanced by the air pressure inside the can, quickly crushes it into a shapeless mass of metal (see Figure 1*b*). The process begins within less than a minute and requires several minutes for completion. It may be accelerated by pouring cold water on the can.

Quick-Crunch Method

A variation of this demonstration is illustrated in Figure 2. Heat an aluminum soft drink can containing a small amount of water, holding it by tongs over the Bunsen-burner flame until water vapor escapes profusely from the opening. (Good heating is essential for the proper effect.) Quickly invert the top of the can in a container of water and—crunch! This is a faster version of the preceding demonstration. (I first saw the phenomenon demonstrated by those wizards from VMI, Rae Carpenter and Dick Mannix.)

The cool water quickly condenses the steam in the can until the partial vacuum creates a "fountain effect" inside the can. This is evidenced by the draining of water from the can when you raise it from the container of water. (Not only does the soda can present a contrast, but also duplicating fluid cans are becoming hard to find with the increased use of electrostatic copier machines.)

I now use a pair of ice cube tongs instead of the beaker tongs shown in Figure 2. A first grade class to whom I was speaking accused me of crunching the soda can with the tongs. Even so, I always have a second can available and ask, "Would you like to see that again?" They always do. To preclude any tong-crunching doubt, this time I drop the inverted can into the water. ■

Submitted by George Kauffman, Department of Chemistry, California State University, Fresno, California 93740. Soda can variation added by column editor.

Share your favorite demonstration with the readers of the *Journal*. Mail a description of it (or them) to me, Jerry D. Wilson, Department of Science and Mathematics, Lander College, Greenwood, SC 29646; (803)229-8386.

Figure 1. (*a*) Before. (*b*) After.

Figure 2. Soda can quick-crunch variation.

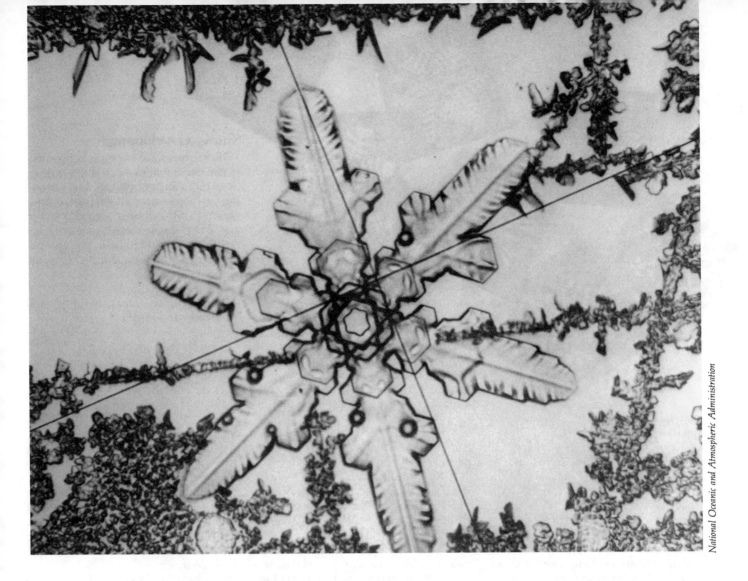

National Oceanic and Atmospheric Administration

Permanent Snowflakes

By Larry Schafer

How would your students like to save the snowflakes that fall in January to study under their microscopes in June? And how would they like to project the snowflake images on the walls of their warm classrooms? Almost everyone has made prints *in* the snow. But how many have made prints *of* the snow?

If you teach in a climate where it gets below freezing and snows—no matter how infrequently—and if you're willing to make some simple preparations, you and your students can turn snowflakes into fossils for study and permanent pleasure.

To Catch a Flake

Start by gathering a few inexpensive and readily available items. You'll need

- a can of clear plastic spray (Krylon™ #1303 works well, but others do too—experiment with what's available)
- 5–10 clean glass microscope slides (or 5 × 5-cm squares) of thin window glass—pieces bigger than this will not fit into slide projectors
- 4–6 spring-type wooden clothespins—these will grab the slides more securely than do plastic pins
- 4 clean plastic bags with wire ties
- 2 sturdy cardboard boxes with tops (liquor store cartons are fine)

Recommended but not absolutely essential is a freezer or a refrigerator with a freezing compartment, either of which is handy for keeping materials cold in preparation for on-the-spot snowflake catching.

Now, you'll be ready to settle back and wait for a snowfall and below freezing temperatures.

Getting Cool

Be prepared with backups in case you or your students accidentally drop an entire bag of pins or slides in the snow—put half the glass slides in each of the two plastic bags and close them securely with the wire ties. Similarly divide, bag, and secure the clothespins.

An hour or so before you want to begin preserving snowflakes, cool the

(Above) A photograph of a magnified snowflake displaying the hexagonal structure characteristic of these ice crystals.

Earth at Hand

materials you'll be using to below freezing (the colder the better). Place the can of Krylon, the two bags of slides, and the two bags of clothespins in a freezer (or, if the temperature outside is below freezing, in a covered box outside in the shade). Also cool the two cartons out-of doors in a calm, sheltered place.

Now, with both materials and air temperature below freezing and snow falling, you are ready to save the snow.

1. Take materials out of the freezer. Use gloves as you grasp the tops of the bags so that you don't warm up the slides and pins.

2. Hurry outside and put this equipment in one of the boxes, which will serve as your arctic supply cabinet.

3. With gloved hands, use a clothespin to pick up a slide, grabbing it on one side only. Quickly shut both bags to avoid getting falling snow on the unused slides and pins.

4. Holding the slide up with the clothespin, spray it with the cold plastic. (Be sure not to fog over the slide by touching it or breathing on it.) Allow the excess plastic to drain off one of the corners.

5. Then, hold the slide flat to catch the snowflakes.

6. After a few flakes have landed on the slide, put it in the other box, plastic

side up. Keep the lid of the box closed to prevent further flakes from landing on the slides.

7. Repeat this process until you have a number of exposed slides resting in the closed box.

8. Return the plastic spray can, pins, and unused slides to the freezing compartment.

After more than an hour outside in the cold, closed box, the plastic on the slides hardens, and the water disappears through it to leave little hollow snowflake replicas. *(Sublimation* is the name of the process by which the snow disappears. This process is related to evaporation except that, in sublimation, a solid turns to a vapor without first becoming a liquid. Dry ice, as it disappears, is also undergoing sublimation.)

When you think the slides may be ready, take one indoors to make sure all the snow has disappeared (undergone sublimation). Otherwise, when you bring the carton of slides inside, the snow will melt and cause the snowflake fossils to collapse.

If your trial slide passes the test, the next step will be looking at the snowflake images either under a microscope or as projected on a screen by your slide projector without its tray. *Be careful not to touch the surface of the slides and crush the plastic snowflakes.*

Also, do not stack the slides. Store them, covered against the dust, in flat boxes.

Larry Schafer is an associate professor of science teaching at Syracuse (New York) University. Snowflake designs taken from **Handbook of Designs and Devices** *by Clarence P. Hornung.*

Snowy Developments

Buy some diazo paper, a light-sensitive material that makes positive images, at a store stocking blueprint, engineering, architecture, or art supplies. (Dietzgen 241 BSF™ white is a good brand to ask for.) Because diazo paper does not react very quickly to light, you can use it in a dimly lit room for a short time without exposing it.

You'll also need

- 2 clean, empty coffee cans (one pound or larger and identical in size)
- enough clean stones or marbles to cover the bottom of 1 can
- nonsudsing household ammonia*
- masking or cellophane tape

To develop the slowflake images

1. Cut the diazo paper into approximately 13 x 18-cm sheets. Work in a dimly lit room, and try to keep the paper, in general, away from the light as much as possible; in particular, keep the yellow light-sensitive side down.

2. Either wrap the sheets in black paper, or put them in a light-proof box with the yellow sides down.

3. Set up a 35-mm slide projector about 1 m from a white wall. Take off the slide tray.

4. Now, insert a snowflake slide into the projector and focus the image of the snowflakes on the wall. Make sure the lamp is on high to get the brightest possible image.

5. Put the stones or marbles in one coffee can, and add a little ammonia—enough to cover the bottom of the can but not the top of the stones or marbles. Place the empty can upside down on top of the first to contain the fumes as you work.

---

*Be careful that no one breathes the ammonia fumes.

6. At this point, shut off the lights to darken the room. Absolute darkness is unnecessary: you should be able to see what you are doing. Retrieve one of the diazo sheets from its box, and carefully shut the lid. Then tape the sheet, yellow side out, to the wall so that the snowflake images fall on the paper. Make sure the paper lies flat.

7. Expose the sheet until just after the yellow disappears. If you leave the sheet for too short a period, no images will appear; if you leave it too long, the overdeveloped paper will become entirely blue.

Don't worry if a little yellow remains around the edges when you take the paper down.

8. Carefully curl the exposed sheet inside the upper coffee can so that the exposed once-yellow side faces inward to catch the fumes. Again, taking care that neither you nor your students breathe the vapors, return the top can to its former upside-down position above the can with stones and ammonia. The fumes will turn the yellow or shaded areas blue.

Once all the snowflake images have turned blue, the paper will not fade in normal light.

Lessons in the Snow

Use the images you've created to make some comparisons and see some contrasts:

- Examine the symmetry of snowflakes. Are there lines that could cut a snowflake into two identical halves?
- Make paper cutouts that look exactly like real snowflakes.
- Group or sort cutouts of snowflake prints first according to shape and then according to size. Do snowflakes that have nearly the same shape also have nearly the same size?

- Do all snowflakes have the same number of sides?
- Are snowflakes caught at the same time about the same in shape? In size?
- How do snowflakes gathered on windy days compare in size and shape with those from calm days? How do those from heavy falls compare with those from light ones?

Plowing Deeper

Your students may want to move from their lovely, lacy snowflake images to the library to begin to try to do some forecasting along some of the following lines.

- Can you predict from the shape and size of snowflakes the kind of weather that is on its way?
- Is there a relationship between the size and shape of the flakes and the air temperature? The amount of cloud cover—sunny, partly cloudy, or cloudy? The presence of wind or calm?

Snowflakes have many lessons to teach both on their surfaces and in their depths. Be ready to teach your class snowflaky science the next time snow covers your cold winter landscape.

Resources

Vogl, Sonia. (1982, January). Outdoor science in winter. *S&C, 19*(4), 15–17.

Williams, Terry Tempest, and Major, Ted. (1984). *Secret language of snow.* San Francisco: Sierra Pantheon.

This "flake" is made of four crystals rather than one. These crystals, which are in the form of bullet-shaped hexagonal columns, are attached at the bottom.

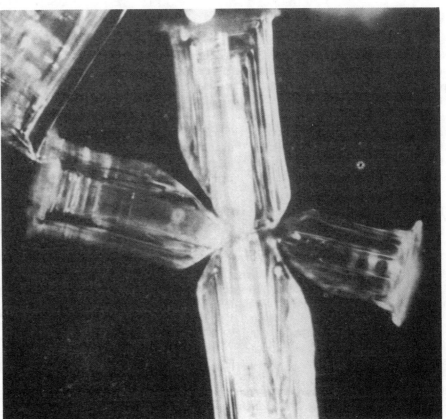

Snowy Science

As a child growing up in Michigan, I learned to love the winter. I don't remember the days as being cold or windy, I just remember the many hours spent outdoors in the snow, making snowmen, snowballs, and snow forts and discovering first-hand the properties of this wonderful substance. Without realizing it I learned a lot of science. Your students can also learn science from the snow if you do some planning.

Record snowfalls are a surefire way to introduce the topic of snow. According to the *Guinness Book of World Records* the biggest recorded snow in a 24-hour period was the 193 centimeters (cm) that fell at Silver Lake, Colorado in 1921. The record for a single snowstorm, 480 cm, was registered at Mount Shasta Ski Bowl, California. And over 4,300 cm fell on Mt. Rainier, Washington, during a 1971-72 period, for the greatest snowfall recorded in a 12-month period. Have your students try to guess each of these records. If they can come to a consensus, have them measure out the estimated depths to get a better feel for them. Then tell them the true records. With interest piqued, students can begin outdoor activity.

Snowflakes

Catching, observing, and preserving snowflakes can be fascinating. The first step is to put pieces of dark construction paper or cloth into a freezer several hours before going outside. When it snows, each of your students can catch a few flakes on the dark surface and examine them with a magnifying glass. Have them draw and describe several crystals. Is each one unique? Are there different types of crystals? Get students to research this question using one of several sources that classify crystal types into different categories, for instance, Snow and Ice. Students can also measure snowflake diameter with a ruler calibrated in millimeters.

Snowflakes can be permanently preserved on glass slides. Have slides, pieces of cardboard, and a can of clear plastic or lacquer spray chilling in the freezer. When the time comes to collect the flakes, get each of your students to put a slide on a cardboard piece and quickly take it outside. Carrying the slide this way will keep the snowflakes from being melted by warm hands. After collecting one or more flakes on a slide, a student should gently spray the slide with lacquer. It can then be allowed to dry outside for a few hours in a spot where it is protected from the snow. When the slides dry, students can bring them inside to observe the snowflakes with a magnifier or a microscope. The flakes will melt when warmed but the lacquer will preserve their shape and general outline.

Measuring Snow

There are many ways to measure snow. A meter stick inserted into the snow will tell the depth. Or you can make a snow gauge from a straight-sided tin can or jar by marking off depths in centimeters from

the bottom of the can to the top. Put the gauge outside before a snowfall and check it when the snowstorm is over. If a number of gauges are put out, students will be able to compare readings from gauges at different points around the school. How might wind affect the measured amount in any one place? How can students determine the rate per hour at which the snow fell? You may wish to ask them to chart the snow accumulation over time.

Collected snow can also be melted to determine the volume ratio of snow to water. How much water does 150 cm of snow yield? Get students to collect several samples of new-fallen snow of equal volume, being careful not to compact the snow as they collect it. Does the ratio stay constant from one sample to another? What factors might cause changes in this ratio? Is the ratio constant from one snowstorm to the next? Is it the same for new snow and older snow?

Snow and Temperature

How cold is snow? Does the temperature vary depending on whether the snow is in shade or in sun? Whether the snow is clean or dirty? Snow can be colored by sprinkling tempera paint on it from a salt shaker. Will the color of the snow affect the snow's temperature on a sunny day? Will one color of snow melt faster than another? Investigations of this type can give your students many opportunities to practice process skills.

Snowballs can also be the focus of experiments. Students can compare the melting time of different size snowballs. How do packed snow, fluffy snow, and ice cubes compare? If you don't want to wait for a whole snowball to melt, get your students to measure the amount of melt water released during a certain time period, perhaps five minutes.

Drifting and Snow Fences

If you are in an area that gets a lot of snow, investigating drifts can be an interesting activity. To get some hints about how drifts are formed, your students should observe the movements of snow while it is being blown into drifts. Obtain pieces of plywood about 1/2 meters square, and nail sticks to them so that the squares can be propped up on the ground. Put one square on edge so that the prevailing wind blows directly at it. Put another at a 45° angle to the wind and yet another at a 90° angle. Get students to describe how the drifts differ around these boards. Do they form on the front or the back of the boards? How far does the drift form from the board? How might the size of the board affect the drift?

Older snowdrifts can be used to study the processes of snow compaction and aging. Find a drift formed by several different snowfalls. Carefully cut into it with a shovel to expose a vertical cross section. Get students to see how many layers of snow can be counted. Are the crystals of the same consistency in each layer? Ask students if they can find evidence of thaws. Can they speculate as to how long each layer was exposed to the elements? (An estimate of this can be made by comparing the amount of dirt on the top of every layer.)

Students may also be interested in

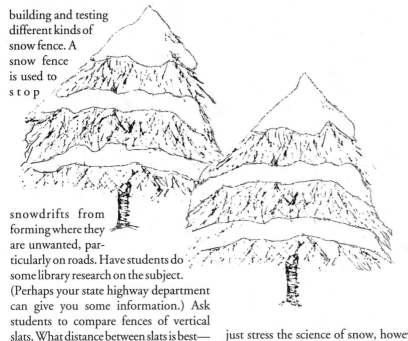

building and testing different kinds of snow fence. A snow fence is used to stop snowdrifts from forming where they are unwanted, particularly on roads. Have students do some library research on the subject. (Perhaps your state highway department can give you some information.) Ask students to compare fences of vertical slats. What distance between slats is best—one, five, or ten centimeters? They can also compare vertical with horizontal slats or with some other design of fence. What characteristics do well designed snow fences have? Why aren't snow fences made of solid sheets of wood?

Early adolescents, indeed most children, are naturally fascinated by snow. Using this natural interest can turn the outdoors into a winter laboratory. Don't just stress the science of snow, however. Getting students interested in the beauty of snow, and of winter, can produce an important change in attitude in some students. Extending science into art with snow sculpting and painting or into literature with snow poetry can prove to be the missing ingredient for many.

Michael Padilla
University of Georgia
Athens, GA

The Salt Oscillator

Small increases in the amount of dissolved solids in water can have profound effects an biological, chemical, and mechanical systems dependent on specific densities, for example, a submarine or a fish. In a 1971 issue of *Scientific American*, Seelye Martin writes about his discovery which he calls a salt oscillator. Briefly, dyed salt water in a paper cup is placed in a beaker of fresh water. A pinhole in the cup's bottom lets the heavier salt water flow downward into the fresh, but suddenly the flow abruptly ceases, and fresh water mysteriously begins to flow into the paper cup. After several seconds, the salt water once again begins to drop, and then, like clockwork, the process reverses itself. One of Martin's oscillators ran for four days before equalizing.

Many variations of this lab are possible. A salt fountain can be created, salty water can be replaced with very cold water or hot water can stand in for fresh water. Students will have some understanding of the general principle fairly quickly though I would avoid a long digression on the actual switching mechanism. They will be interested in playing around with different fluids such as syrups and detergents.

Mark S. Wiley
George School
Newtown, PA

Oleander

Gone With the Wind?

Demonstrate the process of erosion by planting a marked pole in the beach. Any place on the beach will do—erosion, alas, is widespread. Have the students revisit the pole two or three times at regular intervals, noting on each occasion the lowest mark visible. With luck their scientific project will not have become some beach-comber's firewood and the results will be quickly apparent.

Next, back in the classroom, show your students how wind and vegetation work on soil by setting up a simulated beach (sandpile) in a pan with shallow sides. Aim a hand-held blow dryer at the "beach" and watch it "erode." Now cover the "beach" with plants (simply *laying* the plants down will protect the sand; you don't even need to anchor them) and blow again.

Your students will quickly see that the sand builds up around the plants. Thus they'll realize that plants prevent erosion. Now they'll be ready to examine other threats to the wetlands and suggest other ways of helping.

Venetia R. Butler and
Ellen M. Roach
Savannah-Chatham School System
Savannah, GA

Earth in Space

Photographing the Night Sky (Without a Telescope)

by Roger L. Scott

Why use store-bought slides of constellations, comets, and meteors when you can take great shots in your own backyard with a 35-mm camera?

Many people think you can only photograph the night sky if you use telescopes or telephoto lenses. In fact, however, you can take beautiful celestial photographs with an ordinary 35-mm camera mounted on a sturdy tripod. Color film can record vividly all the stars you can see as well as many others too faint for your unaided eye.

Telescopes and telephoto lenses enlarge the image of a distant object. When that distant object is in the sky, however, the rotation of the Earth (including the telescope or telephoto lens and the person looking through it) interferes with getting a good, detailed photograph of the celestial object. Telescopes and telephoto lenses must be attached to motor-driven mounts to compensate for the effect of the turning Earth. Otherwise the object of interest quickly drifts out of the field of view.

Luckily for the amateur celestial photographer, the image produced by the "normal" lens supplied with most 35-mm cameras is small enough so that the Earth's rotation will hardly distort photos taken during short-duration exposures. You do not need an expensive motor-driven mount for exposures of only a few seconds.

If your purpose is not to produce a detailed photograph of a single heavenly body, but to get sharp photos of what your students have studied only on star charts, a short-exposure photograph of the backyard variety is often just the ticket.

Which film should I choose?
Although a number of black-and-white

Object	Kodachrome 64 Ektachrome 64	Ektachrome 200 Ektachrome 400
constellations	lens wide open 20–40 sec.	f/2.5 20 sec.
star trails	f/5.6 several minutes	f/8 several minutes
bright comets	lens wide open 20–40 sec.	f/2.5 20 sec.
aurorae	lens wide open 20–40 sec.	f/2.5 20 sec.
meteor showers	f/5.6 several minutes	f/8 several minutes

Figure 1. Examples of exposures and lens settings and exposures for night-sky photography with a 50-mm focal length lens.

and color films can be used for successful night-sky photography, it is difficult to obtain good prints of sky exposures from a commercial processor. Therefore, unless you are skilled in darkroom printing techniques, it is best to use color slide film. Since this film develops directly into a positive color transparency, no printing is required. Small hand-held slide viewers are inexpensive and easy to use, and slides are certainly better for classroom use than prints.

To avoid confusing color slide films with color print films, remember that slide films have the word "chrome" in their names, while color print films are designated by the word "color." Examples are Kodachrome 64 (for slides) and Kodacolor-X (for prints).

Why choose color film for the black-and-white night sky? Because your color slides will reveal what your students can't see with their unaided eyes. Young, hot stars will appear blue or white; older, cooler stars will be red or orange; and all the stars will be tiny, brilliantly hued jewels when projected on your classroom screen.

In general, "faster" (more light-sensitive) films are best for night-sky photography. Relative light sensitivity is expressed as an ASA or ISO number printed on the film box. Commonly available color slide films range from ASA 25 to ASA 400 and higher. The higher the number, the greater the sensitivity of the film to light. This sensitivity is calibrated for exposures of snapshot duration—that is, fractions of a second. Fast films tend to produce a somewhat grainier image, resulting in a slight loss of detail.

For successful night-sky photography, use film with an ASA of at least 64. Suitable films include Kodachrome 64 (ASA 64), Ektachrome 64 (ASA 64), Ektachrome 200 (ASA 200), and Ektachrome 400 (ASA 400). If your sky is very dark, try using a fast film such as Ektachrome 400. Use slower films, such as Ektachrome 64 and 200, in the city to avoid overexposing the background sky.

Roger L. Scott is an associate professor of physics and astronomy and director of the Planetarium and Observatory at Ball State University, Muncie, IN 47306.

—Roger L. Scott

Figure 2. Orion the Hunter. Reproduced from a Kodachrome 64 slide, which was exposed for 40 seconds at f/1.4. Star colors are very vivid on the original slide; the red giant Betelgeuse appears bright orange and the Orion Nebula pink (due to hydrogen α emission).

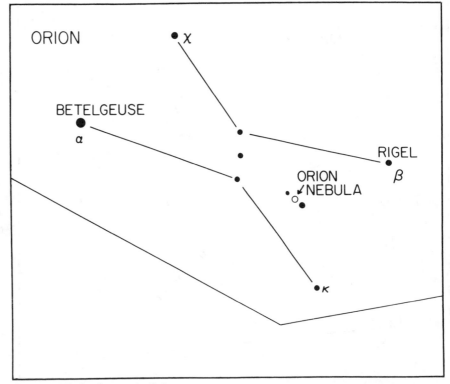

Figure 3. The bright stars in Orion.

—Roger L. Scott

Figure 4. Canis Major the Great Dog. Reproduced from a Kodachrome 64 slide, which was exposed for 30 seconds at f/1.4. Canis Major is Orion's great hunting dog. Sirius is the brightest star visible in the night sky.

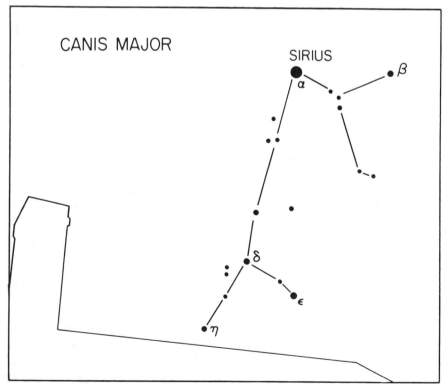

Figure 5. The prominent stars in Canis Major.

The lens as an eye

Night-sky photography without a motor-driven mount is best done with a nontelephoto lens—one that does not magnify. If your camera has a built-in lens, it will be nontelephoto. If your camera is designed for interchangeable lenses, you will want to use one with an approximate focal length of 50 mm or less.

The amount of light entering the camera lens is controlled by an iris diaphragm similar to the iris of the human eye. The iris of the camera is wide open when the lens is set to the smallest f-stop number; it is barely open at the highest f-stop number. You can "stop down" (close up) the lens in a series of standard steps. Typical settings range from f/1.4 (wide open) to f/16 (minimum aperture).

Modern lenses will usually perform reasonably well wide open, but the sharpest image is produced when they are closed down a few stops. The photographs accompanying this article were taken with the iris wide open in order to capture the maximum amount of light in the shortest possible time. If you examine the images carefully, you will see some slight distortion in stars near the edge of the photo. Try your first night-sky exposures with a variety of f-stop settings, starting at wide open and closing down one f-stop between successive exposures. This will let you see if opening your lens to maximum aperture produces objectionable distortion in the star images.

Your next steps are loading the camera with the appropriate film and mounting the camera on a sturdy tripod. Attach a cable release (available at any camera store) to the shutter release control. This allows you to open and to close the shutter without actually touching the camera and possibly jarring it. Set the shutter speed control to "B," or "Bulb," so that the shutter remains open as long as the cable release plunger is pressed in. Older cameras may have a "T," or "Time," shutter speed setting. When the shutter is set to T, pressing and releasing the cable release plunger will open the shutter. The plunger must then be pressed and released again to close the shutter. Be sure that the lens is set at the desired f-stop and that the focus is set to infinity (∞). Use a lens hood, if you have one, to avoid problems from stray light.

108

Ready to shoot?

Once your camera is loaded with film and correctly adjusted, you are ready for night-sky photography. Simply locate the object you want to shoot in the viewfinder, and open the shutter for as long as you think necessary. Experiment with different lengths of time. The sky doesn't have to be totally dark for these shots—but be careful that the background light in the city does not overwhelm the object you're trying to photograph. Sample different f-stops to see what works best. Include some familiar object, such as your house, in the field of view to lend scale and add to the effect. I've suggested some sample exposure conditions in Figure 1. This table is intended only as a rough guide; experiment on your own to find the settings that work best for your purposes.

Modern astronomers have divided the sky into 88 regions, or constellations; some of these are always available as photographic subjects on a clear night. Astronomers and casual stargazers alike recognize the constellations by the ancient patterns formed by the bright stars visible to the naked eye.

Over the course of a calendar year, you can easily photograph all the bright constellations visible from your locale. You will find these slides especially valuable for classroom use. Students quickly can learn to recognize a constellation's star pattern and the types of stars—from white dwarfs to red giants. Star charts, such as the monthly charts published in *Astronomy, Sky & Telescope,* and *Science and Children,* are helpful aids for identifying constellations.

Figures 2 through 5 show the winter constellations of Orion the Hunter and his great dog Canis Major. These prints were made from Kodachrome 64 slides. The stars have trailed slightly because of the Earth's rotation. When you are not using a moving mount, longer exposures such as these will result in oblong star images or even trails. But in some instances this can be a desirable, rather than a detrimental, effect.

A 50-mm lens will produce noticeable star trails when exposures exceed 20 seconds. To get good photos of star trails, you must leave the shutter open for several minutes, perhaps even an hour. Remember that for such long exposures, you must stop down the lens to keep light from the background sky from becoming too bright and from making it harder to identify the star trails. A star-trail exposure centered on Polaris, the North Star, will demonstrate dramatically the Earth's rotation. All stars will trail in concentric arcs centered on the North Celestial Pole, the point in the sky above Earth's north pole.

You can also photograph bright comets, aurorae, meteor showers, and more by this method. Bright comets are easy subjects. The same exposure conditions that work well for constellations are also suitable for comets. Figure 6 is a 30-second exposure of Comet Bennett, originally taken on Ektachrome HS, or "High Speed," a film no longer made (ASA 160). Note how the antenna and building lend scale to the photograph.

The Aurora Borealis, or Northern Lights, is caused by charged particles that are ejected from the Sun and then interact with the Earth's upper atmosphere. You can often see aurorae a day or so after a major solar flare. Look for reports in the news media about solar outbursts. You can photograph aurorae using the same exposure conditions that work well for constellations and comets.

Showered with meteors

Several times each year the Earth passes through the debris-strewn orbit of a comet; the particles that burn up in our atmosphere create what is known as a meteor shower. Though the comet fragments are moving in parallel orbits about the Sun, perspective makes the resulting meteors appear to diverge from a point, or radiant, in the sky. An average shower may produce 20 to 50 meteors per hour, all streaking from the vicinity of the radiant. Meteor showers may be photographed using the same exposure conditions used for star trails. Annual showers, which may last for 2 or 3 days, are named after the constellation containing the radiant.

Meteors should be photographed with your lens stopped down, as indicated in Figure 1, in order to avoid overexposing the background sky. If you are away from city lights, with a totally dark sky, you may open up your lens.

The modern word "planet" comes from an ancient Greek word meaning "wanderer," because as the planets orbit the Sun they appear to drift relative to the background stars. Try verifying this yourself. Pick out one of the bright planets, such as Mars, Jupiter,

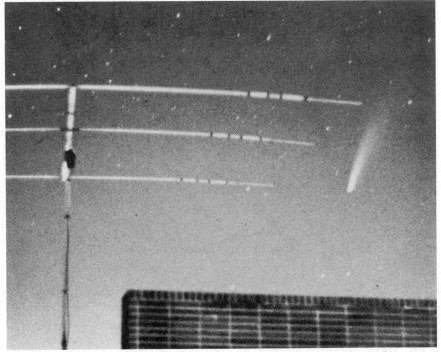

—*Roger L. Scott*

Figure 6. Comet Bennett. Reproduced from the original Ektachrome HS slide (ASA 160), which was exposed for 30 seconds at f/1.4. The building and antenna were included to lend scale to the photograph.

or Saturn, in the night sky. Watch it for a few nights; you will see it moving across the sky while the stars around it appear to stay in place. For an excellent classroom project, you might take a series of slides showing the motion of a planet through a bright constellation. Since the bright planets appear as "stars" in the night sky, you can photograph them in the same manner as constellations.

Although I wrote this article with a 35-mm camera in mind, any camera capable of taking a time exposure will work to some degree. I have even used a Polaroid camera to produce acceptable black-and-white prints of constellations and comets.

It is extremely easy to take beautiful, educational photographs of the night sky without a telescope. With the simplest 35-mm camera you can use night photography to teach your students everything from how Earth's rotation affects what we see in the night sky, to what color a young star is, to how the planets orbit the Sun. So start those shutters snapping! ■

For further reading
1. *Astronomy* magazine, $21 per year, published monthly by AstroMedia Corp., 411 E. Mason St., PO Box 92788, Milwaukee, WI 53202. *Astronomy,* which includes a monthly sky calendar, is well known for its beautifully illustrated articles on astronomical photography.
2. Norton, A. *Norton's Star Atlas,* 17th ed. Cambridge, Mass.: Sky Publishing Corp., 1978. Classic reprint, available for $18.95 from Sky Publishing Corp., 49 Bay State Rd., Cambridge, MA 02238, contains detailed star charts and lists of interesting astronomical objects. The book is invaluable for identifying constellations.
3. Paul, H. *Outer Space Photography for the Amateur,* 4th ed. New York: American Photographic Book Publishing Co., 1979. Available for $10.95 from Sky Publishing Corp., this book is the standard reference for night-sky photographers and is crammed with examples of astronomical photography and technical know-how.
4. "Sky Calendar," published monthly by Abrams Planetarium, Michigan State University, East Lansing, MI 48824. "Sky Calendar" is a very useful monthly star chart, showing the positions of the bright constellations and planets. "Sky Calendar" has a current annual subscription rate of $5. ("Sky Calendar" also appears each month in *Science and Children.*)
5. *Sky & Telescope* magazine, $18 per year, published monthly by Sky Publishing Corp. Though written for a more sophisticated reader than *Astronomy,* *Sky & Telescope* always contains articles of interest to the night-sky photographer. A monthly sky calendar is included.

Why the North Star Doesn't Appear to Move

Why does the North Star appear to remain stationary while all the other stars in the sky appear to move? Students and adults ask this question when they see the apparent motion of the stars in the nighttime sky. The answer that the North Star is nearly over the axis, or turning point of Earth, is not satisfactory to many.

I have found it helpful to compare the sky's apparent movement to the turning of a merry-go-round, and other stars represent the horses. If you stand in the center with the North Star, you only turn in your place, but if you ride on any other star (horse), you move around the center as the merry-go-round turns.

Through the use of this analogy stu-

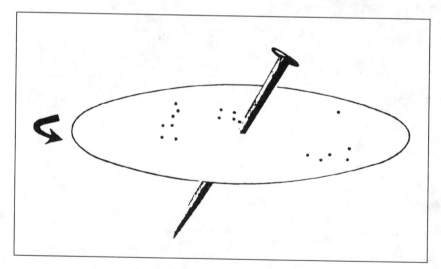

dents understand the idea that the central point can remain stationary while the other points revolve around it.

Illustrate this effect by making a star merry-go-round:
• Cut out a circle marked with stars.
• Place the North Star in the center.
• Push a thin nail through the North Star.
• Holding the nail stationary, turn the circle of stars.

Julie Cassen
Charlotte-Mecklinburg Schools
Charlotte, NC

Early Adolescence

NASA

This column brings special attention to the concerns and needs of the middle/junior high teacher. Send ideas to the column editor, Michael Leyden, Professor of Science Education, Eastern Illinois University, Charleston 61920.

Celestial Meetings— Synodic Periods

It's exam time in your seventh-grade astronomy class, and students sit, heads bent, scribbling furiously on their papers. You can almost hear the sigh of relief when they reach the two questions they *knew* you'd include:
• How long does it take the various planets to orbit the Sun?
• What distance separates each planet from the Sun?
Primed for these questions, students rapidly pencil in their answers, the ones they memorized from their textbooks last night, the ones they'll probably forget tomorrow.

But what if students had learned to derive the numbers needed to answer these questions rather than simply memorizing them? The logic and geometry needed for such tasks are surprisingly simple, and the concepts involved are not entirely intimidating. What's more, using this approach to planetary distances, you'll have the satisfaction of knowing that students have learned to use reason, not just rote learning, and that five, maybe ten, years from now (instead of five or ten minutes), they'll still be able to answer these questions. This month I'll illustrate how the periods of the planet were determined and next month explain the geometry of computing planetary distances.

The Planets' Stately Movements

It's impossible to determine how long it takes a planet to orbit the Sun simply by observation. After all, we're looking at the movement of other planets from a planet that is also in motion. Instead, the length of time it takes a planet to orbit the Sun must be calculated from other observations about the planet's orbit relative to the Earth's movement.

Figure 1 shows the position of a planet

(P-1) and the Earth (E-1) plotted against the background of stars. One year later (measuring by Earth time, of course) the Earth has returned to its original position (E-1 and E-2). In the same amount of time, however, the slower moving planet has only advanced to point P-2. As more time passes, the Earth will catch up to the planet (E-3 and P-3), and the two will again be in opposition.

At this point some definitions are in order. First, those slower moving planets whose orbits are outside the Earth's are called *superior planets*, and the time it takes the Earth to "lap" one of these slower moving bodies is called their *synodic period* (from the Latin *synodus*, meaning *meeting*). Inferior planets, on the other hand, are those (like Mercury and Venus) whose orbit is closer to the Sun than Earth's is. Their synodic period occurs when one of them laps the Earth in this orbital race. In other words, a synodic period measures the interval between the start of the race and the time of the first lapping, and superior plants move one less lap than the Earth in their synodic periods; inferior planets, one lap more.

In Motion 'Round the Sun

With this appreciation of the planetary race, students should now be ready to learn to calculate a planet's *sidereal period*, or the time it takes the planet to orbit the Sun. To simplify the mathematical analysis, let's begin with an imaginary inferior planet whose synodic period is five years. By our definitions, this planet made six orbits in the time it took the Earth to make five. Resorting to some simple computations, we figure that the planet moved 2160° (six orbits of 360° each), a rate of 36° per month (2160°/60 months). This means it would take the planet just 10 months to orbit the Sun (360°/36° per month).

Now we will apply the same logic to Mercury. The arithmetic is not quite as neat, but the process is similar. It takes the planet Mercury 116 days to lap the Earth. In this time, the Earth covers 116/365ths of its orbit, or (since the orbit is 360°) about 116°. By definition, the faster moving Mercury will have traveled one more lap than the Earth, or 476° (360° + 116°). This means that Mercury travels at the rate of 4.1°(476° ÷ 116 days) and would complete an orbit of the Sun in about 88 days (360°/4.1° per day). The figure of 88 days is what many children memorize, though they have no idea of its origins.

Let's skip to the other side of the Earth's orbit and compute the sidereal period of an imaginary superior planet. If its synodic period is five Earth years, it makes four orbits during these 60 months, one less orbit than the Earth. Since it traveled 1440°(four orbits of 360° each) in 60 months, it must have been traveling at the rate of 24° per month (1440°/60 months). This would make its sidereal period 15 months long (360°/24° per month).

Again, let's apply the same logic to a real superior planet, in this case, Mars. Its synodic period is observed to be 780 days. In that time, the Earth makes 780/365 orbits, or 2.14 revolutions around the Sun. Thus, the Earth traveled 770° (2.14 × 360°), and Mars made one less revolution to total 410°. This means Mars travels about 0.526° per day (410°/770 days) and would orbit the Sun in 684 days (360°/0.526°per day).

Celestial Calculations

These questions will help students review the process of determining sidereal periods.
- What is the planet's synodic period?
- Is the planet moving slower or faster than Earth?
- How many degrees did the planet move in its synodic period?
- How fast is the planet moving in degrees per day or month?
- How long would it take the planet to move 360°?

Figure 2 shows the synodic periods that are observed from Earth and the sidereal periods also calculated from the Earth. Have students use the questions given above to see if they can calculate the sidereal periods on their own.

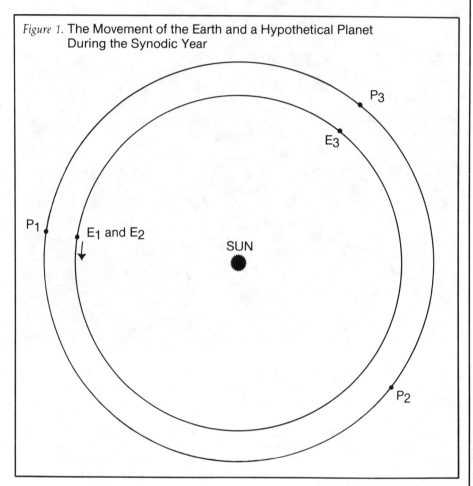

Figure 1. **The Movement of the Earth and a Hypothetical Planet During the Synodic Year**

Figure 2. **Steps in Figuring Sidereal Periods**

Planet	Synodic Period	Degrees Earth Moves* in That Time Period	Degrees Planet Moves* in That Time Period	Degrees Planet Moves/Day	Sidereal Period	
Mercury	116 days	116°	476°	4.1°	88	days
Venus	584 days				225	days
Earth	————	————	————		365	days
Mars	780 days	770°	410°	0.526°	687	days
Jupiter	399 days				11.9	years
Saturn	378 days				29.5	years
Uranus	370 days				84	years
Neptune	367 days				165	years
Pluto	367 days				248	years

*For simplicity, assume the Earth moves 1° per day

Deriving the sidereal peiods for the planets represents the true essence of science: thoughtful analysis of information derived from careful observation.

Next month your students will have a chance for more of the same as they learn another astronomical technique—calculating planetary distances.

National Science Teachers Association

Set Your Class in Celestial Motion

Teaching the three Rs of astronomy—

rotation, revolution, and retrograde motion.

by Dorothy Lane Lamark

The inherent problem with teaching astronomy is that classes are held during the day and most astronomical events happen at night. To get around this problem, I developed an astronomy simulation activity that takes place right in the classroom. The activity introduces my students to difficult concepts that usually require a planetarium to demonstrate—rotation, revolution, and retrograde motion.

Before actually beginning the activity, you may wish to discuss some of the early myths and theories that explained what people observed in the sky. You may be surprised by the number of myths and theories your students are already familiar with. Make sure to include beliefs such as those held by the ancient Egyptians, Romans, and Greeks, particularly those that explained the apparent wandering of the planets as the independent meandering of the gods. Later, during the activity, the students will better understand this belief when they observe this "wandering" for themselves. If possible, have your students explain what logic, observations, or other data supported the myth, theory, or idea being discussed.

You should begin the activity near the end of a class period, when you have about 15 minutes to spare. Tell your students to imagine that what they see on the walls and ceiling of their classroom is really the night sky and that they must construct a map of the sky on paper. They should start by drawing a circle that encompasses the entire sheet. The next step is to make marks along the circumference that correspond to the hour positions on a clock.

When the circles are finished, the class as a group, should choose 12 "constellations." The constellations are classroom objects situated around the perimeter of the room at each clock position. Have each student label the constellations on his or her map. Don't forget to give the constellations official-sounding names; for example, the American flag at twelve o'clock (the front of the room) might be "Americus Flaggus," and the outlet at six o'clock (the back of the room) could be "Electronica." Students usually enjoy this part of the activity. As the bell rings, stress the importance of bringing the sky maps to class tomorrow.

Creating your classroom universe

Before class starts the next day, prepare the room for the activity by clearing the center of the room of all furniture. When rearranging the room, make sure that you don't inadvertently obscure any of the constellations. Next, place a large light source on a cart or table at the center of the room. Tape down the electrical cord to prevent accidents. Before the students enter, turn off all of the lights except for the one in the center of the room. This dramatic effect should get your students' attention and prepare them for something different. Tell the students that they will need their sky maps, a pen or pencil, and something to rest the map on, such as a book. All other materials should be put away. Have the students face the light and form a tight circle around it. Tell them that they each represent the Earth and the light represents the Sun.

Now that you have a captive audience of "Earths," introduce the class to celestial motion. Ask your students how they can tell a star from a planet. Do they know that the planets are among the first observable bodies in the night sky, or more importantly, that the planets appear to wander if observed over a period of time? Be sure to stress that this wandering cannot be noticed from night to night. The sky must be observed over many nights depending upon the planet being observed, your geographical location, and the time of year.

To further explain and demonstrate this point, I initiate a dialogue with my students. I start off by saying, "Imagine that there is a little man living on the end of your nose, standing on the tip looking down at your feet. What time is it for that little man?"

Some students will respond by looking at the clock and telling you what time it is. Eventually, someone will figure out that it is noon for the man, since the students are facing the light, and the Sun is therefore directly over the man's head.

The next thing I say is "Make it midnight for the little man on your nose." Most students know to turn their backs on the light. I ask them to silently note what constellations they see from this position. Then I instruct them to turn until it is noon again for the little man on their noses, and I ask them two questions: "How much time has gone by?" and "What constellations can the little man see?"

Students usually handle the first question well, answering 24 hours or 1 day. The second one gives them more problems. Some students will begin to name constellations. Do not accept any of these responses. Eventually someone will realize that since it is noon for their "little men," the men cannot see any constellations.

Now I ask, "What is the real definition of a day?" I expect students to answer that a day is the amount of time required for the Earth to make one turn on its axis, or one rotation. I follow up with, "Make it dawn for the little man on your nose." The students generally realize that dawn will be halfway between midnight and noon. What they may not know is which "halfway" to choose. Explain with the help of a globe or diagram that they must rotate so that their right shoulder faces the Sun.

The next command is the place where the whole demonstration could fall apart. "Now that we are all facing the same direction, make 6 months go by." Students should move around the circle until they are 180° from where they started. But in order for anyone to move, everyone must move. In groups lacking leadership, you may have to say, "Ready, go!" or "One, two, three, walk!" Don't be upset if students get a little silly during this part of the activity. It really does look funny. Soon the students will be too busy to laugh, so let them get the giggles out of their systems.

After asking the students to define revolution and year, I tell them to "Make it midnight for the little man. Does he see the same constellations he saw 6 months ago? Why or why not?" Students usually know that they are seeing different constellations, but they may need you to point out that

Dorothy Lane Lamark is an earth science teacher who wrote this article while teaching at Merrimack High School in Merrimack, N.H. She can now be reached at 1 George St., Andover, MA 01810.

this is because they are viewing different parts of the sky. A natural follow-up question is, "Why are the constellations different during winter and summer?"

Once the students understand the seasonal changes in the positions of the constellations, we move on to the planets. After telling students to get their sky maps ready, I say, "I am going to represent the planet Mars. You all still represent Earth. Where should I stand and how should I move to be a good model for Mars?" Students usually know that you should be outside their circle, since Mars is farther from the Sun than the Earth is, and that you should move more slowly than they do.

Next, I stand at the chosen point and say, "Plot my position in the sky by writing the number 1 where I appear to you right now, with respect to the constellations." To John, I may appear to be near the constellation "Fire Extingue," but to Alice and Keith I may appear to be halfway between "Fire Extingue" and "Electronica." Since they should have nearly the same point of view as their neighbors, I ask them to check their papers to make sure their plotting is correct.

Expect confusion at this point. At first, students have a hard time grasping the idea that there is not one correct answer, and that no other student will have the exact same answer they do.

After students have my first position plotted, I give them the following instructions: "Move three steps (part of your year), while I move one step in the same direction. Mark my new position with the number 2. Take three more steps, and mark my new position with the number 3 . . ." We continue with this "plodding and plotting" until all the students have returned to or gone past their starting points.

During this part of the activity, you can expect to see some interesting expressions on the faces of your students. Just as the planets appear to

move "backwards" in the sky as the Earth passes them, you will appear to move backwards to your students when they are passing you. Many students will be confused by your apparent backward shift in position, and comments such as "Hey . . . wait a minute!" and "Are you sure you moved?" are common. If you work with the perplexed students and reassure them that they have not made a mistake, they should soon discover what's going on themselves. To finish up, I ask the students to connect the plotted positions on their sky maps. We then discuss the maps, until students can articulate their own definition of retrograde motion.

After this activity has had a chance to sink in overnight (or perhaps fade away completely), you should reinforce the idea of apparent motion by mapping Mars on a real sky map or by demonstrating parallax. A good way to demonstrate parallax, the apparent movement of objects caused by changes in point of view, is to conduct the following simple activity. Have students hold their index finger at arm's length and line it up with an object in the room while keeping one eye closed. Without moving anything, have them switch viewing eyes. Ask them to draw comparisons between this motion and the retrograde motion they observed the day before.

Wrapping up the activity with a further discussion of current and ancient astronomical knowledge should help reinforce the concepts that have been introduced. Your students should now have a greater understanding of why the Greeks and Romans thought that the planet gods wandered the sky of their own volition. Today, astronomers know that this wandering is just retrograde motion. And after your students complete this activity, they'll be one up on the ancients too. ∎

Count Your Lucky Stars

With this astronomy activity, the sky's the limit.

By Gary Tomlinson and Adela Beckman

ASTRONOMY BUFFS AND others alike know that counting stars is similar to counting hairs on a live Kodiak bear. Where do you start and how do you get the bear to hold still? With this activity your students will tame the bear—observe those elusive, twinkling stars and estimate their number.

To Wish Upon a Star Count

Undoubtedly, a person wishing to determine the number of stars that can be seen from the Earth would find it quite a chore, not to say impossible, to count each star visible with the naked eye. Aside from the necessary period of adjustment to seeing in the dark, it's difficult to keep track of the stars you've counted and those you haven't. To complicate matters further, the rotating Earth puts a new slant on the same scene; many stars appear to rise in the east and set in the west right before your very eyes. When you count stars, you need to allow for this effect of the Earth's movement. Also,

as you can see only about one-half of the sky at any one time, your star-counting will fall short by half as well.

With all these considerations, how could a teacher ever count stars with a class? Consider this simple analogy: If we had to determine the number of bricks in a brick wall, we could count each individual brick, or we could count the number of bricks in one square meter and estimate the total by

(Right) Constructing the sky window. (Below) How many bricks in the wall?

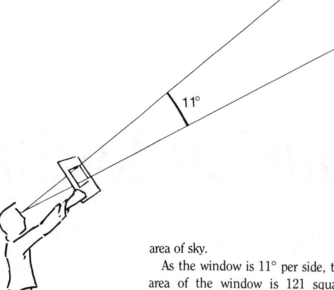

measuring the wall in square meters.

The same type of procedure also applies in estimating the number of stars in the sky. We could determine the number of stars in a certain area of the sky and then calculate the number of areas it would take to cover the entire sky. We would, however, have to devise a way to look at and measure a particular area of the sky. And unlike the square meters of the brick wall, the size of the sky can be measured in square degrees (a unit applicable for measuring small areas of it).

To measure a particular area of the sky, we need something conveniently available—an arm and a hand. The length of the human hand and the length of the arm are in approximately the same proportion during one's life span (Nelson, 1979). When the hand is held perpendicular to the line of vision at arm's length, the angle formed is about 11° (see above). Thus, if everyone could make a square window, each side the length of his hand, and hold it extended in the same manner, we would all be looking at about the same

area of sky.

As the window is 11° per side, the area of the window is 121 square degrees; if we know how many square degrees there are in the sky, we could then divide that by the area of our sky window to determine the number of windows needed to cover the entire sky. Since the sky appears to be a sphere surrounding the Earth, we can do our calculations based on the number of square degrees in a sphere. By checking reference books we find there are approximately 41,253 square degrees in a sphere (Abell, 1969). Thus, mathematical calculations tell us that the number of square degrees in the sky divided by the number of square degrees in the window equals the number of windows per sky.

$$\frac{41,253 \text{ sq. deg./sky}}{121 \text{ sq. deg./window}} = 341 \text{ windows/sky}$$

Making a Sky Window

Now you are ready to help students make sky windows. Make sure you check the weather forecast before you begin the activity—a clear sky for a few days is a must as students will be anxious to estimate the number of stars in the sky with their new sky windows.

To make one, each student will need

- a 25 x 25-cm piece of cardboard or a file folder (it must be at least 10 cm larger than the length of the hand),
- a pair of scissors,
- and masking tape.

Have each student measure the length of his hand. Then cut a square window in the center of the piece of the cardboard the same size as the measured hand length. Show students how to hold the sky window at arm's length and let them practice with it. If you have a room made of cinder blocks or bricks, have students stand the same distance from the wall and measure how many blocks are visible within the window. Have students frame the area with masking tape and measure the length and width (should be the same) of the area they viewed. Once done, have students compare results with classmates.

Next, review area calculation with the class (length x width = area). From their estimates have them calculate how many blocks or bricks make up the wall. Ask a student to count the number of bricks in the wall to check the accuracy of the estimates. Again, mathematical calculation tells us that the area of the wall divided by the area of the window equals the number of windows that will cover the wall. As well, the number of bricks in the window multiplied by the number of windows that will cover the wall equals the number of bricks in the wall.

Now, adapt the procedure to counting the number of stars in the sky. On a clear night have students count the stars in their sky windows. Aiming high in the sky will prevent horizon haze from interfering with the count.

Back in the classroom, take an average of the numbers and multiply the average by 341 (the number of sky windows in the entire sky). The answer should come out to be five or six thousand. This is the number of stars visible with the unaided eye in both the Northern and Southern hemispheres (Pickering, 1958). Since you only look at one-half of the sky at any one time, the students will actually have viewed only about 3,000 stars in the sky on their observation night.

For Further Investigation

Have one group of students take a sample of stars in the city and another group take a sample of stars in the

All set for a star count.

country. Have them look in the same direction (east, for example) and at the same height, to reduce the number of variables. They should focus on a bright star in the sky and keep that star in the same position in the sky window. Average both sets of results and compare. The country sample should be larger.

Or, you may want each student to take two star samplings. One sample should be taken immediately upon coming out of a brightly lit house, and the second should be taken after the student has remained in the dark for 20 to 30 minutes. Compare and see how the eyes have adjusted.

As students begin exploring the possibilities of the sky, further their learning with a thought-provoking

Check the weather forecast before beginning the activity. Students will be anxious to try out their sky windows.

question. Why is it impossible (assuming you are not at the Earth's equator) for a person to see all of the five to six thousand different stars during the course of the year? Explain to the class that some stars are visible all year long to one hemisphere or the other; they never rise or set, they only circle around the celestial pole. We never see these so-called circumpolar stars in the hemisphere opposite our own. Students will enjoy discovering the answers and may come up with intriguing questions themselves in their research.

There are countless variations of astronomy activities to try and to experience. An easy-to-make, easy-to-use sky window is only the first step to a stellar science unit.

Resources

Abell, G. (1969). *Exploration of the universe* (2nd ed.). New York: Holt, Rinehart, and Winston.

Henning, L.A. (1981). Mathematics concepts in the planetarium. *The Planetarian, 10*(4), 14-15.

Joseph, J.M., and Lippincott, S.L. (1969). Let's count the stars. *Science and Children, 6*(5), 65-63.

Nelson, W. (1979). *Textbook of pediatrics* (11th ed.). Philadelphia: W.B. Sanders.

Pickering, J.S. (1958). *1001 questions answered about astronomy.* New York: Dodd, Mead.

GARY TOMLINSON is a curator of the Public Museum of Grand Rapids in Michigan. ADELA BECKMAN teaches sixth grade at the C.A. Frost School, also in Grand Rapids. Photographs courtesy of the authors.

Earth at Hand

117

Starry Archaeology

———— Darrel B. Hoff ————

Although most students begin astronomy by studying the visible sky, they are not likely to learn much about the origin of our present constellations. The subject, however, is a fascinating one that lends itself to an open-ended investigation involving observing, inferring, and model building. The activity might be described as starry archaeology, and it calls on students to reconstruct the society that named the 40 most ancient constellations visible in our hemisphere from information given by the names themselves. The activity is challenging enough to be used with students in college, but it will also work well with children as young as fifth graders. Before beginning the activity, you will want to give the students a little background about constellation names.

A Carved Marble Globe

Everyone knows that the visible stars in the sky belong to constellations (often represented by pictorial outlines). Our modern constellations were codified by a commission of the International Astronomical Union during the 1920s and are 88 in number(2),* but some 40 of the constellation names are very ancient. Exactly how ancient no one knows.

They were already in existence by 370 B.C., when Eudoxus, a contemporary of Plato and Aristotle, is believed to have transported a celestial globe out of Egypt. This globe, marble with bas-relief figures of 40-odd constellations carved on it and about 65 centimeters in diameter, was later incorporated into a statue of Atlas supporting the globe. It is called the Farnese Atlas and can still be seen at the National Museum in Naples, Italy.(1) The fame of the globe was spread in antiquity by a Greek narrative poem, "The Phenomena," which was later translated into Latin by Cicero, as well as by the Greek astronomer Ptolemy, who gave directions for

*See References.

Darrel B. Hoff is a professor of astronomy and science education at the University of Northern Iowa, Cedar Falls. Celestial maps by Guy Ottewell from his Astronomical Calendar 1983, *available from the Department of Physics, Furman University, Greenville, South Carolina, for $10*

constructing celestial globes in his astronomical work, *The Almagest*, and who made several himself. In time, the constellations named on the globe became a permanent part of the sky lore of the West. This familiarity, however, tells us nothing about the origins of the names.

The Unknown Namers

We can assume that the people who named the constellations called them after objects, animals, and supernatural or human figures that were familiar and important to them. So the names can provide evidence for reconstructing the culture of these people, whoever they were.

Tell your students that they are going to do an archaeological reconstruction based on words rather than bones, stones, or bits of pottery. Divide students into teams of four or five each and pick a recorder-reporter for each group. Then give them the names of the 48 constellations with their English translations (See Box.)

Ask them to assume that the names all came from a single society in one era (this may not be the case, of course) and ask them to infer all they can about this society on the basis of the names of the constellations.

Some students may know the stories, but encourage them not to worry too much about details. It's enough to think of Hercules as a superhero and Andromeda as a princess in danger of being devoured by a sea monster.

Suggest that, by looking at the constellation names, they will find clues about the following aspects of that culture:

geographic location (climate, proximity to sea, topography)
level of technology (tools, mechanical devices)
religion (nature of deity or deities)
family structure
political system
common occupations
arts
mathematical sophistication

Mention that absence of evidence *is* evidence. For example, what kinds of animals appear and what kinds do not? There are large land animals (e.g., the bear) but no elephants or *polar* bears. There are no hippopotamuses or tigers. What does this say about the possible location of the region in which the people lived? Encourage students to speculate about names that don't seem to offer much information. (What might

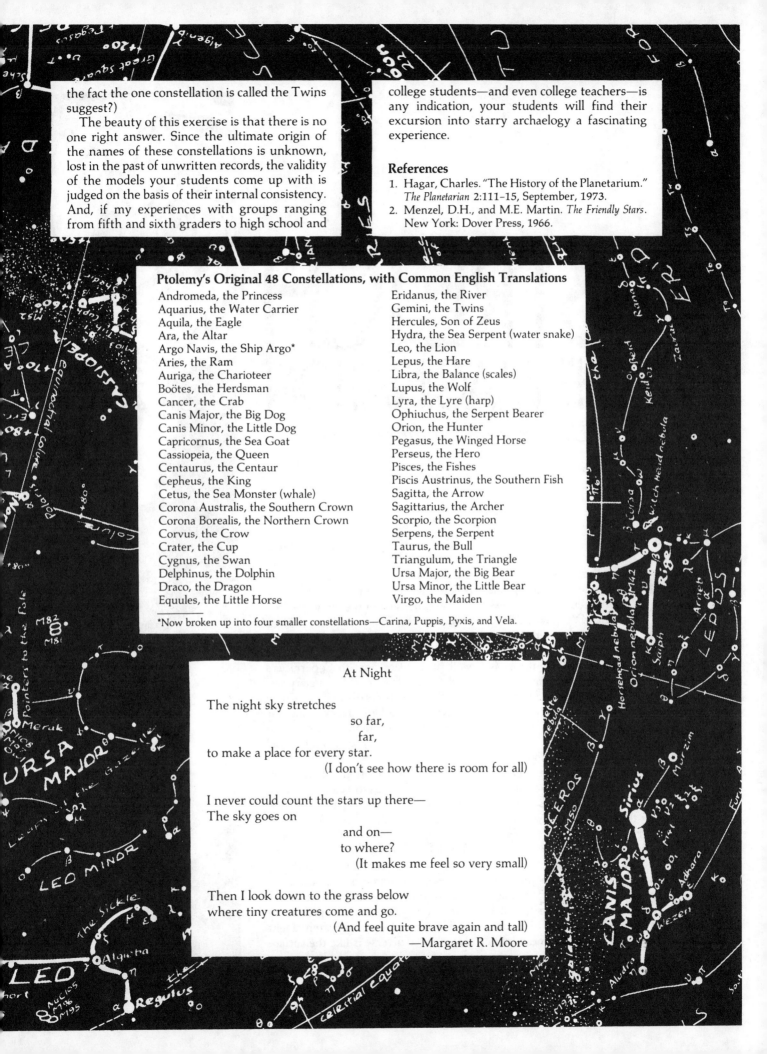

the fact the one constellation is called the Twins suggest?)

The beauty of this exercise is that there is no one right answer. Since the ultimate origin of the names of these constellations is unknown, lost in the past of unwritten records, the validity of the models your students come up with is judged on the basis of their internal consistency. And, if my experiences with groups ranging from fifth and sixth graders to high school and college students—and even college teachers—is any indication, your students will find their excursion into starry archaelogy a fascinating experience.

References

1. Hagar, Charles. "The History of the Planetarium." *The Planetarian* 2:111–15, September, 1973.
2. Menzel, D.H., and M.E. Martin. *The Friendly Stars.* New York: Dover Press, 1966.

Ptolemy's Original 48 Constellations, with Common English Translations

Andromeda, the Princess	Eridanus, the River
Aquarius, the Water Carrier	Gemini, the Twins
Aquila, the Eagle	Hercules, Son of Zeus
Ara, the Altar	Hydra, the Sea Serpent (water snake)
Argo Navis, the Ship Argo*	Leo, the Lion
Aries, the Ram	Lepus, the Hare
Auriga, the Charioteer	Libra, the Balance (scales)
Boötes, the Herdsman	Lupus, the Wolf
Cancer, the Crab	Lyra, the Lyre (harp)
Canis Major, the Big Dog	Ophiuchus, the Serpent Bearer
Canis Minor, the Little Dog	Orion, the Hunter
Capricornus, the Sea Goat	Pegasus, the Winged Horse
Cassiopeia, the Queen	Perseus, the Hero
Centaurus, the Centaur	Pisces, the Fishes
Cepheus, the King	Piscis Austrinus, the Southern Fish
Cetus, the Sea Monster (whale)	Sagitta, the Arrow
Corona Australis, the Southern Crown	Sagittarius, the Archer
Corona Borealis, the Northern Crown	Scorpio, the Scorpion
Corvus, the Crow	Serpens, the Serpent
Crater, the Cup	Taurus, the Bull
Cygnus, the Swan	Triangulum, the Triangle
Delphinus, the Dolphin	Ursa Major, the Big Bear
Draco, the Dragon	Ursa Minor, the Little Bear
Equules, the Little Horse	Virgo, the Maiden

*Now broken up into four smaller constellations—Carina, Puppis, Pyxis, and Vela.

At Night

The night sky stretches
        so far,
        far,
to make a place for every star.
          (I don't see how there is room for all)

I never could count the stars up there—
The sky goes on
        and on—
        to where?
          (It makes me feel so very small)

Then I look down to the grass below
where tiny creatures come and go.
          (And feel quite brave again and tall)
          —Margaret R. Moore

A STRETCH
OF THE IMAGINATION
The Universe on a rubber band

by Alan Lightman

Scientists claim that the Universe is expanding, but without any center of expansion. Yet how can there be expansion without any center? This basic feature of the Big Bang theory is difficult to fathom and difficult to teach to our students. What astronomers actually see through their telescopes is that the galaxies are speeding away from Earth in every direction they look. But shouldn't that outward flight of galaxies mean that we are the center of some local explosion? Why should we think that the galaxies are flying away from *every* location in space, that space itself is expanding?

Putting aside these conceptual questions, many people know nothing of modern cosmology. Recently, Jon Miller of Northern Illinois University and I conducted a national survey of astronomical literacy in the American public. We found that only 24% of adults know that the galaxies are in moving away from us.

Also surprising to us was the finding that only 37% of adults believe that the Sun will burn out, and only 55% believe that the Sun is a star.

The expansion of the Universe, first observed and interpreted in 1929, is arguably the most important astronomical discovery of the century; it may well fit in place next to Copernicus's model of a Sun-centered planetary system. Because of this discovery, we now believe that the Universe was once far denser and hotter than it is today, and that the Universe had a beginning, and it may have an end.

As with the demotion of Earth from the center of things to a circling planet, the discovery of the expansion of the Universe has changed our view.

EXPANDING ON THE SUBJECT

In the last two years, I have developed and used a classroom demonstration that conveys both an intuitive and a quantitative grasp of the expansion of the Universe. An analogy often taught is that the Universe is like the surface of an expanding balloon. If dots are painted on the balloon, each representing a galaxy, then from the view of any

National Science Teachers Association

one dot all the other dots appear to move outward when the balloon is inflated.

While this analogy indeed shows how a material can expand without any center of expansion, it always raises questions about the meaning of the *inside* of the balloon. Furthermore, it does not show why the analogy follows from the observed motions of galaxies, which are seen only from our one dot, the Milky Way.

In my demonstration, the Universe is represented by a rubber band. Cut a long, sturdy rubber band, so you have a

Before

your students

begin this

experiment,

review the crucial

observations

of astronomer

Edwin Hubble.

single length of rubber at least eight inches long, and draw ink marks on the rubber band (see Figure 1). Each ink mark represents a galaxy. Although the locations of the ink marks can be arbitrary, it is best for measuring purposes to place your ink marks at regular intervals, about 1 inch apart, when the rubber band is unstretched. Place a ruler next to the rubber band.

Before your students begin this experiment, review the crucial observa-

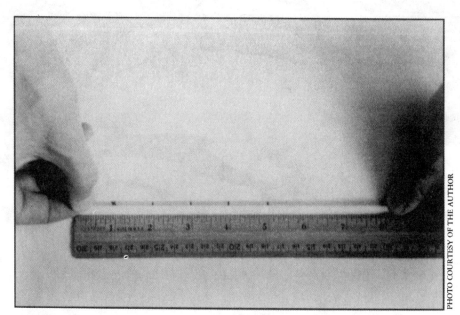

FIGURE 1. *The rubber-band Universe in its initial position, beside a ruler for measurement. Each ink mark represents a galaxy, and are 1 inch apart. The cross is the Milky Way, all distances will be measured relative to it.*

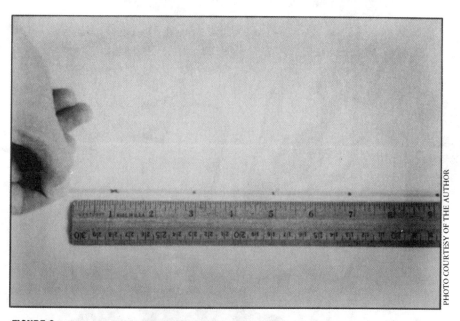

FIGURE 2. *The rubber band has been stretched so that all ink marks are 2 inches apart. The ink mark initially 1 inch to the right of the Milky Way is now 2 inches away; it has moved 1 inch. The ink mark initially 2 inches to the right is now 4 inches away; it has moved 2 inches, and so on. Distance moved is proportional to initial distance.*

National Science Teachers Association

tions of astronomer Edwin Hubble in 1929. Hubble recorded not only that galaxies were moving radially outward from Earth, but also that the speed of each galaxy was proportional to its distance away. That is, a galaxy twice as far away as another galaxy is moving outward twice as fast. It was this last quantitative result, later called "Hubble's Law," that supported the concept of a Universe in uniform expansion, without a center of expansion. The following experiment will illustrate this point.

GETTING UNDERWAY

To begin, hold each end of your rubber band and stretch it. Declare one ink mark to be the Milky Way and measure all distances and speeds relative to it. It is best to keep the Milky Way ink mark at a fixed point on the ruler, say at the 1 inch mark. Upon stretching, you will find that each ink mark moves a distance proportional to its initial distance from the Milky Way ink mark. For example, when the ink mark initially 1 inch away moves to 2 inches away, the ink mark initially 2 inches away moves to 4 inches away. The first ink mark has moved 1 inch while the second has moved 2 inches (Figure 2). Since these increased distances are accomplished in the same period of time, the second ink mark moves twice as fast as the first. In other words, speed is proportional to distance. In fact, any uniformly stretching material produces the law that speed is proportional to distance. If the material is not uniformly stretching, with some sections stretching at a greater rate than others, then speed is no longer proportional to distance. Conversely, the proportion of speed to distance traveled means that the material is stretching at a uniform rate.

It is also easy to demonstrate that the expansion has no center or privi-

References and Further Reading
Lightman, A.P. and Miller, J.D. 1989 "Contemporary Cosmological Beliefs." *Social Studies of Science* 19: 127–136. 1977.

Weinberg, S. *The First Three Minutes* New York: Basic Books.

leged position. Any ink mark can be chosen as the Milky Way, with all distances and speeds measured relative to it, and the result is the same: the other

It is easy to demonstrate that the expansion has no center or privileged position.

ink marks move away from it with speeds proportional to their distances. No ink mark is special; the view is the same for all. If speed were not proportional to distance, then the view would not be the same for every ink mark.

The rubber-band experiment can be done either by individual students, or demonstrated to a large group in a lecture format. In the latter case, attach a paper clip to each ink mark, for visibility, and draw a ruling with inch marks

on a plastic transparency. In this way, the experiment can be projected onto a large screen using an overhead projector.

SOME HISTORICAL NOTES

A factor leading to Isaac Newton's belief that the Universe was static was his inability to imagine a cosmos in motion without any center of motion. In a famous letter to theologian Richard Bentley in 1692, Newton argued that if the Universe were globally expanding or contracting, there would have to be a center of motion; but since matter scattered uniformly through infinite space does not define a center, the Universe must be static. Albert Einstein, in his cosmological model of 1917, simply assumed that the Universe was static and sought static solutions to his equations. After Edwin Hubble's discovery of 1929, Einstein agreed that the Universe was in a state of uniform expansion. Hubble's law was theoretically predicted in 1927 by the Belgian priest and scientist Georges Lemaitre, who proposed an expanding Universe model.

From the observed rate of expansion—each galaxy will double its distance from us in about 10 billion years—astronomers have estimated that the Universe is about 10 billion years old. This number is remarkably close to the age of terrestrial uranium ore, dated by its radioactivity and mixture with lead. Thus, by two completely different means, one using the movements of distant galaxies and one using rocks underfoot, scientists have arrived at a consistent measure of the age of the Universe.

Alan Lightman is professor of science and writing and a senior lecturer in physics at Massachusetts Institute of Technology, Room 14N-431, Cambridge, MA 02139. He is also a staff member of the Harvard-Smithsonian Center for Astrophysics, in Cambridge.

MODELING THE MILKY WAY

Spreadsheet science

by John C. Whitmer

The Milky Way galaxy can be seen as a faint luminous band of light across the sky on a clear, moonless night. It is a vast system of gas clouds, dust, and about 100 billion star that includes our Sun.) Most, however, are much too distant to be seen clearly with the naked eye. This collection, bound by gravity and rotating slowly around a central region, has a diameter of about 100 000 light years (1 light year = 9.47×10^{12} km), and our solar system is a tiny subsystem about two-thirds of the way out from the center. The statement that light would take 100 000 years to cross the Milky Way conveys little to most people about its size.

To illustrate the dimensions of such a system, teachers must rely on scale models. By using an electronic spreadsheet, you can for models calculate to describe the relative dimensions of our solar system and the Milky Way galaxy. Once programmed, the spreadsheet

FIGURE 1. The spreadsheet used here is QUATTRO (Borland), but any spreadsheet will work. Cells containing formulas are shown at the bottom.

```
        A          B          C          D          E          F          G          H
1                      MODELING THE SOLAR SYSTEM AND GALAXY
2
3          This spreadsheet calculates the dimensions for a scale model of the
4       planets, the sun, the next nearest star and our galaxy.  You choose the
5       scale by entering any real dimension in cell C8 and the corresponding
6       model dimension in cell C9.
7
8          REAL (km): 1391000            PROPORTIONALITY CONSTANT
9          MODEL (m):    0.25            model(m)/real(km) : 1.8E-07
10                                       model(cm)/real(km): 1.8E-05
11                                       model(km)/real(km): 1.8E-10
12      ------------------------------------------------------------------------
13          |        REAL DIMENSIONS         |       MODEL DIMENSIONS
14          |equatorial      distance        |                   distance
15     SOLAR| diameter      from sun         |    diameter       from sun
16     SYSTEM|  (km)          (km)           |     (cm)            (m)
17      ------------------------------------------------------------------------
18     SUN   | 1391000              0        |    25.00            0.0
19     Mercury|   4880      5.79E+07         |     0.09           10.4
20     Venus  |  12100      1.08E+08         |     0.22           19.4
21     Earth  |  12756      1.5E+08          |     0.23           27.0
22     Mars   |   6794      2.28E+08         |     0.12           41.0
23     Jupiter| 143200      7.78E+08         |     2.57          139.0
24     Saturn | 120000      1.43E+09         |     2.16          257.0
25     Uranus |  51800      2.87E+09         |     0.93          515.0
26     Neptune|  49500      4.5E+09          |     0.89          808.8
27     Pluto  |   3000      5.9E+09          |     0.05         1060.4
28      ------------------------------------------------------------------------
29     NEXT NEAREST STAR BEYOND THE SUN (APHA CENTAURI)
30
31         real distance from the sun:         model distance from the sun:
32              4.3   light years
33          4.1E+13   kilometers                  7319   kilometers
34      ------------------------------------------------------------------------
35     MILKY WAY GALAXY
36
37         real diameter of the galaxy:        model diameter of the galaxy:
38           100000   light years
39          9.5E+17   kilometers               1.7E+08   kilometers
40      ------------------------------------------------------------------------
```

Cells containing formulas		
H9: +C9/C8	F18: +B18*H10 copied to cells F19 through	B33: +B32*9.47E + 12
H10: +H9*100	F27	F33: +H11*B33
H11: +H9/1000	H18: +D18*H9 copied to cells H19 through	B39: +B38*9.47E + 12
	H27	F39: +H11*B39

can be used to describe models of various dimensions and scales very easily and quickly.

Spreadsheets are used widely in the business world to organize financial data, however, their flexibility and ease-of-use are equally useful in science classes. Knowing how to use a spreadsheet is more valuable for many students than knowing a programming language. Powerful spreadsheets are now available, often with educational discounts, for less than $50 and many have graphing capabilities. Spreadsheets allow students to spend more time planning and interpreting their work and, consequently, less time on repetitious calculation. Some familiarity with spreadsheets is assumed here. "Spreadsheets Answer What If...?" by Pogge and Lunetta (*The Science Teacher*, 54:46-49, Nov. 1987) includes a brief review of spreadsheet operation. However, the best source of information is the documentation that accompanies your spreadsheet software.

Spreadsheet cells may contain text, numbers, or formulas that can refer to other cells. If two cells contain any corresponding real and model dimensions, another cell can use these to compute the proportionality constant. Then, if a column of cells contains real dimensions, another column can compute model dimensions. The advantage of a spreadsheet is that, once "programmed," all results can be quickly recalculated so that models with different scales can be obtained with little effort.

The spreadsheet in Figure 1 calculates the dimensions of a scale model of the Milky Way including our solar system and the nearest star (other than the Sun), Alpha Centauri, which is actually a group of three closely orbiting stars that appear as one to the naked eye. Data on the planets, Alpha Centauri, and the Milky Way are entered on the left side of the spreadsheet. (Planet data are from the NASA booklet *A Look at the Planets*, U.S. Government Printing Office, 1985. Data on Alpha Centauri and the Milky Way are from "Journey into the Universe through Time and Space," a supplement to the *National Geographic*, 163:704A, June 1983.) The scale is determined by entering any real dimension in cell C8 and its corresponding model dimension in cell C9.

In Figure 1, the Sun (diameter, 1.39 x 10^6 km) is the size of a basketball (diameter, 0.25 m). The proportionality constant and the model dimensions are calculated on the right side of the spreadsheet in cells containing the appropriate formulas. The cells display the answers, although the formulas used are listed at the bottom of Figure 1.

Have your students make a few predictions prior to calculating dimensions of the model. Hold up a basketball and ask how large the Earth would be if the Sun were the size of a basketball? How large and far away would Pluto be? Would the Milky Way galaxy fit in your state?

It is important for students to understand the calculations involved in these

FOR FURTHER READING
Morrison, P., and P. Morrison. 1982. *Powers of Ten, Scientific American Library*. San Francisco. W.H. Freeman.
Author's Note: This striking book, based on a film of the same name, contains 42 full page illustrations, each differing in dimension by a power of 10 from the subatomic level (10^{-16} m) to the edges of the known Universe (10^{25} m). The accompanying text contains a wealth of information about the relative size of things in the Universe and how we measure them.

models so you should review the proportionality concept thoroughly. It may be useful to have each student calculate a proportionality constant and use it to determine one model dimension that then can be compared to the spreadsheet results. A spreadsheet frees students from repetitious calculation, but it should not mask the reasoning behind the calculation.

Your students may be surprised to learn that, relative to the basketball, Pluto, at the edge of the solar system, would be about the size of a small sand grain (Figure 1, cell F27), over a kilometer away from the basketball (cell H27). Alpha Centauri would be thousands of kilometers away (cell F33), and the diameter of the Milky Way (cell F39) would be more than 10 000 times the diameter of the actual Earth (cell B21)! Clearly, we could not physically construct this model, but perhaps a different scale would provide further insight.

Suppose the Milky Way had a diameter in the model equal to the Earth's real diameter? Although we could enter the galaxy diameter in cell C8, and the Earth's diameter in cell C9 numerically, these diameters are already in cells B39

and B21 so we type "+B39" in cell C8 and "+B21*1000" in cell C9 (the 1000 converts kilometers to meters). Instantaneously the new proportionality constant and all the new model dimensions are recalculated and displayed on the right side of the spreadsheet.

Do you think the Sun would be visible in this model? Could the Sun and Alpha Centauri be contained in the same room? If you have a spreadsheet, set it up to help answer these questions. Perhaps some of your students with an interest in computers could "program" this spreadsheet. Once set up, most students should then be able to use it if given a little direction.

Let your students choose the scale. If you have a class-size computer screen or a computer display projector, have one student suggest a scale and then predict one of the model dimensions before the actual calculation is displayed. If a BB represented the Sun, could the solar system fit in your classroom? If the Earth were the size of a hydrogen atom (diameter = 8 x 10^{-11}m), could you fit the Milky Way in the palm of your hand?

Beyond the Milky Way, galaxies are seen in every direction to the limits of the known Universe. Although this spreadsheet could be extended to include neighboring galaxies or those deeper in space, the distances are so great that it is difficult to attach significant meaning to model dimensions. The actual construction of any of these scale models is, of course, not possible. However, by exploring their own questions, students with some visual imagination can begin to grasp the awesome magnitude of interstellar space and become familiar with a very useful software tool in the process.

John C. Whitmer is a professor of chemistry at Western Washington University, Bellingham, WA 98225.

AS SMART AS A FENCEPOST

— Paul H. Joslin —

Once when I was ten and my father was disgusted with something I'd done, he shouted, "You're stupider than a fencepost!" My defensive reply was, "Fenceposts aren't so stupid!" I still believe that. Every elementary school should have a teaching fencepost. Some classes will want to make their posts ceremonial, decorating them with paint or carvings; others will be content to partake of their posts' wisdom without frills.

With two sunny days in any one place and a fencepost or tree, your students will be able to find north if they look for the shortest shadow cast.

The Case of the Wise Fencepost

From a lowly fencepost, children can learn some of the principles of direction and position; of time, time zones, and time keeping; of movements of the Earth and Sun; of equinoxes and solstices; and of seasonal changes. Were your students deprived of compasses, clocks, and calendars, the fencepost could inform them of time, season (even, on occasion, date), and direction.*

A teaching post must be vertical and cast visible shadows on a patch of nearly level ground where the shadows can be marked and recorded throughout the day. Beware of buildings whose shadows may interfere with that of your post (and of surfaces like streets and parking lots on which children should not write or plant markers).

If your class decides on the ceremonial post, this is probably the time to decorate it, perhaps with paint (which must be able to withstand weather), perhaps with totems typical of Native Americans, perhaps with the school's colors and symbol—let your students decide, and then offer appropriate support. You'll also want to supervise any young artists who decide to carve designs into the post and help them choose safe outdoor paints, for instance.

Sherlock Post

On a sunny day, take the class out as early in the morning as possible to mark the end of your post's shadow with a colorful stone or a decorative stick. Then, ask your students to observe where the Sun is in relation to the post's shadow and to another landmark you have selected in advance that stands due north of the fencepost. Ask students also to compare the fencepost's shadow to the position, length, and direction of their own shadows.

Now, let them do some guessing about the future, not only about the relative position of the Earth and Sun but also about shadows they'll be observing later in the day.

• Ask them such questions as whether the post's shadow will be in the same place later. If not, where do they think it will be at, say, noon? At 3 P.M.? How long do they think it will be at those times?

• Pointing at the Sun's location vis-a-vis the designated landmark, ask where in the sky they think the Sun will be at lunchtime or at school-closing time.

• Ask where the shadows would fall if the class came here tomorrow or next week and marked them at exactly the same times. Do your students think the new marks would be in exactly the same places as today's? If not, where do they think tomorrow's shadows would fall?

• Finally, ask them to move backward in time. Where do they think the shadow was at dawn this morning? How long was it then?

When you ask the children these questions, they may feel frustrated at first. How could a *fencepost* tell them the answers? But the point is to prepare them for the fact that this is just the kind of mystery their wise post can solve.

Now, return to the classroom. But,

---

*Any vertical post planted in any state but northern Alaska can provide students with this information. Those parts of Alaska above the Arctic Circle must do without the fencepost's wisdom because, since the Sun never rises at all in the winter solstice, naturally, the post doesn't cast a shadow. (At the North Pole, the Sun doesn't come up for six months at a time.) However, post time is available to guide students in Jerusalem, as Varda Bar shows on pages 16–18 of this month's *S&C*.

Paul H. Joslin is professor of science education at Drake University, Des Moines, Iowa. Artwork by Johanna Vogelsang.

at regular recorded intervals throughout the day (on the hour, for example), send a couple of children out to mark the post's shadow and to note the position of the Sun with regard to the landmark. Repeat the process the next day, using a different kind or color of marker than that originally employed.

Go West, Young Shadow!

A day or two later, take the whole class out to observe the shadow markings. Now, pose some answerable questions:

• Where was the shortest shadow cast? Let your students see that this shadow points at the landmark and inform them that the direction in question is north. (At this point, you can also teach your students where south, east, and west appear.) Then, ask them where the Sun was when the post cast its shadow eastwards.

• Move now to a consideration of the lesson of the longest shadow. Where was it cast? In what direction did it point? At what time was it cast? And where was the Sun then?

Tell the children to look at the location of the earliest shadow of the day and to infer where an earlier shadow would have fallen had they been there to observe it. Then, ask your students to predict where a shadow cast later than the one they are watching would fall.

Again, avoid giving the answers—instead, help students to state their inferences and supporting reasons in the best possible form, encouraging classmates to find support for or to try to refute the hypotheses that emerge.

Stay as neutral as possible.

It's a Clock!

Have your students lay a rope along the shadow markers to form a half ellipse. Lead them to notice that the shortest shadow, if extended, will cut the semi-ellipse into halves and that this extended shadow marks the halfway point of the Sun's journey across the sky and thus divides the day into halves. Show them that the Sun moves in an opposite direction (east to west) from the shadow's path radiating from the post (west to east). See if the fact that the shadow moves in a clockwise motion across the semi-ellipse could help the children make the shadow markings into a crude clock.

Show them that the hourly shadow

markers from the first and second days seem to be in exactly the same positions. (When the children move on to their study of the fencepost as calendar, they'll learn that more accurate measurements over a longer period of time would actually show gradual changes in the daily shadow positions.)

Now extend another length of rope along the short noon shadow in both directions, and explain to the students that this line is a meridian, an imaginary line on the Earth's surface connecting the South and North Poles. (Of course, it will also point directly at your landmark.) In the morning, before the Sun crosses this line, we count the hours as being A.M., for ante (Latin *before*) the north-south meridian. P.M. means post (Latin *past*) the meridian when, in the afternoon, the Sun has passed west of the line. Now the fencepost has taught your pupils the literal meaning of A.M. and P.M.

But the lesson is not over yet. What if the Sun is directly over the meridian? Students will probably know by now that this happens at noon, and they can learn to designate that time properly as 12:00 M., as it should be but is not widely done.

Here, you'll again have to leave the schoolyard for the classroom and temporarily abandon the experimental and intuitive method you've been using to explain why the shortest shadow doesn't fall at exactly 12:00 M. according to local clock time. A map marked with the lines of longitude and the boundaries of the time zones will help here—your school is unlikely to be located precisely upon the line of longitude that determines the standard time for your zone. To estimate the difference between "post" time (or solar time) and that of your students' wristwatches, subtract four minutes from post time for every degree that you are east of the standard meridian in your time zone, and add four minutes for every degree, west.**

If your students wish to be *very* accurate, they must also account for the fact that the Sun will be ahead (up to 14 minutes in the fall) or behind (up to 16 in the spring) depending upon the Earth's orbital position. But, in a society where people go by fenceposts rather than digital watches, standards of time

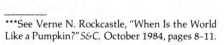

are relaxed, and such refinements may be unnecessary.

It's a Compass!

Your students should now be able to realize that, given two sunny days in any one place and a tree, fencepost, or any perpendicular stick, they can find north (and, therefore, south, east, and west) by looking for the shortest shadow cast. But they can do more than find their way home.

Back in the classroom, you can use the concept of the meridian to teach them about time measurement in a global sense (the time zone map will again be useful) and about navigation.***

It's a Calendar!

Your post and its shadows can, in

**See Marianna Zimmerman, "Time to Build Sundials," *S&C*, November/December 1981, pages 25–26, 35.

***See Verne N. Rockcastle, "When Is the World Like a Pumpkin?" *S&C*, October 1984, pages 8–11.

It's Superpost!

Having introduced the interrelation of time, location, and direction through direct observation of the temporal and spatial interplay between Earth and Sun, shadow and post, your pupils are now ready not only to function as primitives without compasses, clocks, or calendars, but also to take on the refinements of Superpost's lessons with which civilization may decide to confront them.

Elementary, My Dear Fencepost

The shortest shadow cast by the post will tell your students several things: it will show noon local time; the length of the shadow will reflect the season; the end of the shadow will point true North; and, if extended in both directions, it will mark an Earth meridian. Throughout the year, the length of the shortest daily shadow changes; it grows longer in summer and fall and shorter in winter and spring. When the shadow is at its annual shortest, the day is close to June 21, the summer solstice and the first day of summer; it's longest on the winter solstice, around December 22, the first day of winter. The intermediate length shadow between summer and winter marks the autumnal equinox, the first day of autumn (about September 22); that between winter and summer, the spring, or vernal, equinox (about March 20).

addition, help the children see how the Sun changes position from north to south throughout the year. Over several months' observation, your students can watch the shadows lengthen as the Sun sinks lower in the sky when autumn slowly turns into winter and shorten as winter becomes spring.

Around the 20th of each month, have the children mark the shadow lengths at consistent times—say, 8:30 A.M., 10:00 A.M., 12:00 M., 2:00 P.M., and 3:30 P.M.—though earlier or later times will better illustrate changes in the Sun's daily positions in the eastern and western skies. The shadows will lengthen until approximately December 22 (winter solstice) and then gradually shorten. Because of vacation, you will not be able to show your students that the opposite is true in summer, but they will probably be willing by then to take your word for the change.

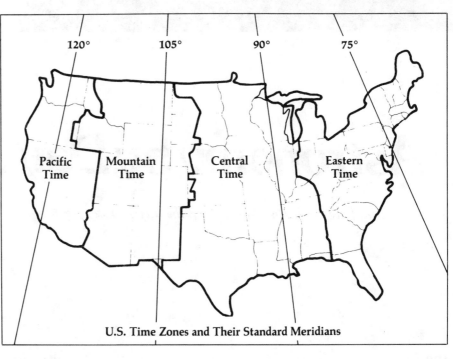

U.S. Time Zones and Their Standard Meridians

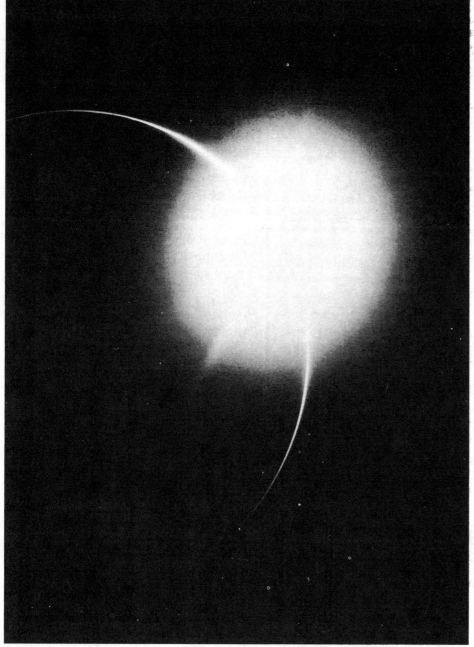

Astronomy by Day

Mapping the liveliest spectacle in the daytime sky—sunspots!

Richard Russo

Making arrangements for evening observation sessions of the Moon, planets, and stars on a clear night—with more than 100 students—is unworkable for the average science teacher. On the other hand, the following daytime observations can involve all your students on a rotating schedule. The program includes regular observations, systematic data recording, and long-term accumulation of information on our "daytime star"—the Sun.

As a result of my participation in the Practical Laboratory Experience Program of the National Science Foundation (NSF), the New York Academy of Science, and Columbia University Teachers College, I prepared this simple experimental project on sunspots for eighth-grade science students. I worked with Daniel P. Hayes, research astronomer at Columbia's Herriman Observatory, whose primary area of research is stellar polarization (in particular, stellar atmospheric models). His research provided a wealth of ideas which formed the background for the sunspot experiment.

To the unaided eye, the Sun appears as the featureless luminary of the daytime sky. When viewed through a telescope, however, it seems to be marked by blemishes—sunspots that are incessantly changing during their lifetimes of several hours to several months. Sunspots appear darker in comparison with other solar regions because they are composed of gases that are up to 1500 K cooler than those of surrounding regions of the photosphere.

Under good conditions, a three-inch telescope resolves approximately 1.5 seconds of arc. This corresponds to a distance of some 1100 km on the Sun's photosphere (a typical sunspot group has a diameter of some 28 000 km). The photosphere

Richard Russo teaches eighth-grade physical science at Montvale Junior High School, Montvale, NJ 07645.

National Science Teachers Association

is the visible layer of the Sun, the layer from which most solar radiation is emitted. The photosphere will provide the data for our investigation.

Observing safety

To observe the Sun with a regular telescope, special precautions must be taken to protect the sensitive retina of the eye from the blinding intensity of the Sun's rays. (Solar filters should never be used, especially those that fit onto the eyepiece.) The only safe way to observe the Sun under magnification is the projection method. Hold or attach a plain piece of paper or oak tag to the telescope behind the eyepiece, to serve as a projection screen. This offers an added advantage: many students can observe the Sun at the same time. (See Figure 1.) Amateur astronomers use the same technique when making solar observations.

Teaming up to scan for sunspots

While working with Daniel Hayes, I realized that one of the strengths of his research program was daily access to the use of instruments. Students could make specific observations over long periods of time. Most professional astronomers, on the other hand, experience down periods in their research, since the major observatories are available to only a limited number of projects at one time.

Our project, like Hayes', has virtually unlimited observing time. Each clear morning, a team of three students set up our three-inch Unitron refractor telescope according to their own projection design which may vary from hand-held to ground mounted. They draw the positions of the sunspots on graph paper corresponding to their positions on the projection screen. The graph paper itself can double as a practical projection screen and a grid to help the team count and coordinate the sunspot groups more readily.

At the end of each month, all observing teams submit their data, which should include an accurate diagram of the positions of solar surface features as well as comments on the numbers, locations, and sizes of sunspots. After each group contrib-

utes its data, each student draws a graph of sunspot numbers as they occurred throughout the month. The graphs are eventually compared with the results published in *Sky & Telescope* magazine.

This program integrates many skills. Students first work in small groups of three, where they make accurate observations using their own projection design. They exchange data, and, therefore, must rely on the ability of their fellow students to collate, graph, and discuss the month's observations. Students can research and discuss such related areas as solar rotation, the life cycle of a sunspot, or the relationship between sunspots and the Earth's weather patterns.

By comparing their results to those of previous years' classes, students may be able to monitor significant changes in the sunspot cycle. E.W. Maunder found evidence that from 1645 to 1715, sunspot activity was low. During this period, referred to as "The Maunder Minimum," Europe experienced a prolonged cold spell.

My class was fascinated to learn that sunspot variations indeed do seem to coincide with changes in the Earth's climate. The specific effects

of these sunspot cycles, lasting 11 and 22 years, on our weather remains a very controversial question. This fact introduces students to a field of current research and stimulates their scientific curiosity.

Before urging you to undertake this experiment, I reemphasize concern for safety. The magnified Sun is lethal! Strongly caution students to stay away from the telescope eyepiece; remove the finder scope to eliminate the temptation. Finding and focusing on the Sun can be both easily and safely accomplished with a properly aligned projection screen.

This investigation can spark an interest in sunspots in particular, in the Sun itself, and in astronomy. Perhaps before long, your students will be out on their own on a clear night, plotting the movements of the Moon, the stars, and the planets. ■

For further reading

1. Eddy, J. "Sunspots." *Natural History*. November, 1976.
2. _____ . "The Maunder Minimum." *Science* 192: 1189-1202, June 18, 1976.
3. Hoyt, D. "Climate Change and Solar Variability." *Weatherwise*. April, 1980.
4. Maran, S. "Counting the Sunspots." *Natural History*. August, 1979.
5. Moche, D. "Observing the Sun with the Projection Method." In *Astronomy-A Self Teaching Guide*. New York: John Wiley & Sons, Inc., 1978.
6. Mullaney, J. "Fine Art of Sunspot Prediction." *Science Digest*. February, 1980.

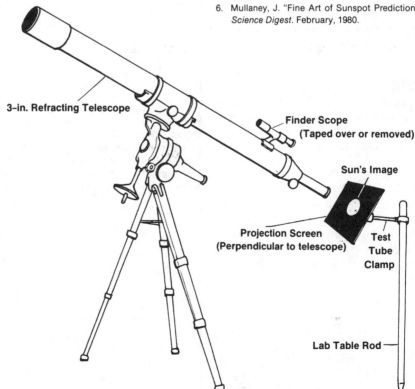

3-in. Refracting Telescope

Finder Scope (Taped over or removed)

Sun's Image

Projection Screen (Perpendicular to telescope)

Test Tube Clamp

Lab Table Rod

Figure 1. The projection method for observing solar features.

Years, Days, and Solar Rays

Try a new way to measure seasonal and diurnal changes in the amount of solar radiation reaching the Earth's surface.

by Richard S. Murphy and John B. Kwasnoski

Many meteorology and earth science classes discuss how the tilting and the daily rotation of the Earth affect the amount of solar radiation that reaches a given area of the Earth's surface (called insolation). The demonstration that teachers usually perform is a *qualitative* one, using a light source and a globe to show how seasonal and diurnal changes in solar radiation occur. But we have discovered an inexpensive technique for *quantitatively* investigating these phenomena.

Setting it up

You will need the usual light source and globe as well as a light meter with a flat photovoltaic sensor. (See Figure 1.) The experimental setup is shown in Figure 2. Note that the light source produces nearly parallel rays of light. We have used a 300-W spotlight, a slide projector, a 300-W floodlight, and an overhead projector and have found all of them to be satisfactory. We do these activities in a darkened room with a dark backdrop behind the

Richard S. Murphy is an assistant professor of physics and John B. Kwasnoski is an associate professor of physics at Western New England College, Springfield, MA 01119.

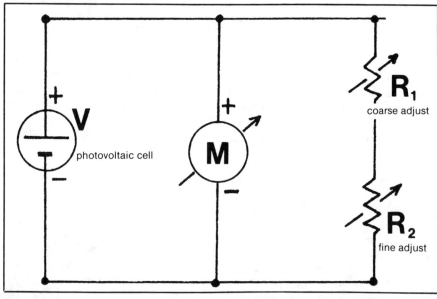

Figure 1. Schematic of the light meter. Parts of the circuit: M = 50-mV galvanometer; R_1 = 1K potentiometer (coarse adjust); R_2 = 200-ohm potentiometer (fine adjust); V = photovoltaic cell (solar cell) with 1 mA-output current capacity. These components or a fully assembled meter can be purchased from Solar Systems International, 27 Greenleaf Dr., Hampden, MA 01036. Many other electronics suppliers offer similar parts.

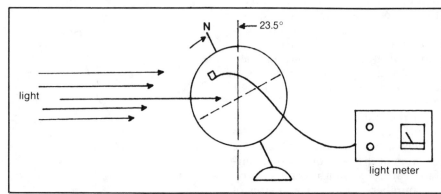

Figure 2. Experimental setup (summer, 23.5° N).

National Science Teachers Association

globe to avoid the nuisance of scattered light.

Start radiating

The tilt of the Earth's axis in relation to the plane of the ecliptic varies with each season. Tilt the globe to a particular position during the year (such as the summer tilt shown in Figure 2). Have a student place the photovoltaic sensor at a particular latitude on the globe (for example, 23.5° north) directly in line with the lamp. Then adjust the sensitivity of the light meter or move the lamp so that the brightest radiation value causes a full-scale deflection of the meter. This corresponds to an insolation level in nature of 1000 W/m².

During a given day, insolation varies with solar altitude and with the length of time between sunrise and sunset. Therefore, to study diurnal radiation levels, have a student place the sensor at a certain latitude and then slowly rotate the globe through 360° (24 hours). Be sure to take note of the time of sunrise and of sunset.

Repeat the procedure for another latitude, or several others, or with the axis of the globe tilted to another season of the year. When you make comparisons of radiation levels, do not move the light source or the globe during the measurements, and do not change the sensitivity of the meter. Typical sets of data for seasonal variations are shown in Figure 3.

Rather than watching you demonstrate the principles behind seasonal and diurnal changes in solar radiation, your students will enjoy the experience of actually measuring the effects themselves and seeing the changes from season to season. Use the data you accumulate as a starting point for a class discussion. ∎

For further reading
1. Byers, H. *General Meteorology*, 4th ed. New York: McGraw-Hill, 1974.
2. Lunde, P. *Solar Thermal Engineering*. New York: John Wiley and Sons, 1980.

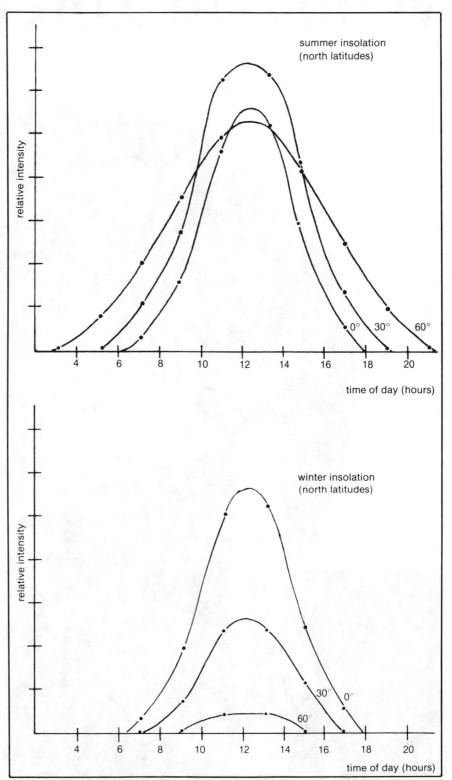

Figure 3. Student data for seasonal variations in the amount of radiation reaching the Earth.

The Earth Is *Round?* Who Are You Kidding?

National Science Teachers Association

Nearly 500 years ago, Columbus proved that the Earth was round. Today's students still don't believe it.

O ne of the first scientific facts we announce to our children is that the Earth is round. That flat, flat ground we walk on, ride on, and play on, stretching endless miles in perfect and reliable flatness, actually wraps around on itself to form a giant ball. Preposterous, yet a scientific fact. For proof, just look at the globe in the classroom or the beautiful photos of Earth taken from space.

By Alan Lightman and Philip Sadler

If we expect young children to believe this story, we're wrong. Studies done in the United States and Israel indicate that as late as the fourth grade almost half the children believe the Earth is flat. Most of the remaining fourth graders who say, "The Earth is a round ball," actually picture a flat part where people live in the *interior* of the ball. Others draw the Earth as a giant pancake or as a curved sky covering a flat ground. These findings run head on against teachers' and parents' perceptions of what their children have learned.

Recently, we asked 65 elementary and middle school teachers to predict how their students would draw the Earth. The teachers greatly overestimated their students' knowledge. For instance, second-grade teachers believed that 95 percent of their students knew the Earth was round when less than five percent of the students actually did.

Clearly, our children are struggling with a contradiction between what they are told about their world and what they see with their own eyes. Such a struggle will inevitably be repeated in other science topics: The world is not always as it seems. The concept of a round Earth presents one of the earliest and most striking examples of this critical lesson in science.

What Looks Flat Might Be Round

The biggest obstacle to accepting a round Earth is that the Earth, indeed, appears flat. To attack this obstacle, we have to show children that something can look flat even when it obviously is not.

In the last year, we have developed and given classroom demonstrations to local first graders using a very large balloon and a toy ship. The demonstration gives young children an intuitive feeling for the shape of the Earth.

We inflate the balloon and have each child, one by one, place her cheek against the balloon, closing the outer eye and looking with the inner eye out over the surface of the balloon. We ask the child to then describe what she sees to the class. Does the horizon look round or flat?

If the balloon is large enough, the surface looks flat and the child says so. We then emphasize to the class that when a person is very close to a large object, it always looks flat. The object might be shaped like a board or a ball or a cigar, but it will always look flat from very close up. In the present case, the children can see that the balloon is round when they are sitting far away from it—just as astronauts can see that the Earth is round from high above it. But when we are sufficiently close to the balloon or the Earth, it appears to be flat.

Ideally, the balloon used for the demonstration should be 10 feet or more in diameter. We used a 12-foot diameter latex weather balloon, which can be bought through local surplus stores, Army-Navy stores, sporting goods stores, and survival stores for about $20 each. The balloon can be inflated in five minutes by reattaching a vacuum cleaner's hose to the exhaust. We inflate the balloon to about 10 feet—to inflate it to 12 feet would stretch it tightly enough to easily pop.

The children should definitely watch as the balloon is inflated. They become enormously excited as this small piece of rubber gets larger and larger, eventually filling up much of the room and forcing them farther and farther back toward the walls of the classroom. For obvious reasons, all sharp objects and edges should be removed from the middle of the room before inflating. Once inflated, the opening nozzle of the balloon may be either tied off or held closed by one of the students during the demonstration.

A Ship on the Horizon

Students should now realize that appearances can be deceptive. The Earth is not *necessarily* flat just because it looks flat. However, this realization still leaves open what the true shape of Earth might be. It could still be flat. And to find out, we need more information.

A second demonstration with the balloon helps. Again, with the child's cheek against the balloon and with her eye closest to the balloon looking at its horizon, we slowly move a toy ship across the surface of the balloon toward the child. As we do this, we ask the child to describe to the class what she sees.

At first, when the ship is on the opposite side of the balloon from the child, she will see nothing. When the ship first appears on the balloon's horizon, the child will see only

As the ship comes over the curved horizon, it is revealed in parts. First the top mast comes in view, followed by the rest. On a flat object, the ship wouldn't be obscured from any distance.

the uppermost part of the ship. As the ship travels closer to her, she will report seeing more and more of the lower parts of the ship. Finally, the ship will be in full view for her. On clear days, this is exactly how sailors see distant ships—little by little, from the top of the mast down. As a matter of fact, Ptolemy, the famous astronomer and geographer, cited this observation as evidence of the Earth's roundness.

By now, the class will begin talking about the curved surface of the balloon as the curved surface of the Earth. But the experiment isn't over.

Using the flat surface of a large table instead of the balloon, we repeat the demonstration with the toy ship. The child reports seeing the *entire* ship at once, no matter how far from her the ship is placed on the table. We emphasize the difference between the two results and the observed evidence of the first.

Naive Theories and Young Minds

Children's deep-seated impression that the Earth is flat is a preconceived notion, or what science educators call a naive theory. Such theories, based on common sense experiences from an early age, are very difficult to give up and can strongly interfere with any learning that challenges them.

Many students, for example, have trouble believing that in a vacuum, a cannon ball and a feather dropped from a tree would fall at the same speed. In the world of experience, air resistance generally causes heavier objects to fall

faster. Even professional scientists can make mistakes by placing too much trust in their personal experiences. Science educators have only recently begun to realize the importance of confronting naive theories and helping students examine their beliefs about the world. Such an examination can begin very early in a child's education.

In this case, all it takes is one very large balloon.

Resources

Champagne, A.B., et al. (1980). Factors influencing the learning of classical mechanics. *American Journal of Physics, 48*(12), 1074–1079.

Nussbaum, J., and Novak, J. (1976). An assessment of children's concepts of the Earth utilizing structured interviews. *Science Educator, 60*(4), 535–550.

Nussbaum, J. (1979). Children's conceptions of the Earth as a cosmic body: A cross-age study. *Science Educator, 63*(1), 83–93.

Sneider, C., and Pulos, S. (1983). Children's cosmographies: Understanding the Earth's shape and gravity. *Science Educator, 67*(2), 205–221.

White, B.Y. (1983). Sources of difficulty in understanding Newtonian dynamics. *Cognitive Science, 7*, 41–65.

Alan Lightman is a research physicist at the Smithsonian Astrophysical Observatory and teaches physics and astronomy at Harvard University. His most recent book is **A Modern Day Yankee in a Connecticut Court.** *Philip Sadler has been a junior high school science teacher. He and his students helped develop the Starlab Portable Planetarium. He is now director of Project STAR, a high school curriculum in astronomy developed at the Harvard-Smithsonian Center for Astrophysics. Photographs by Alan Lightman.*

Shoot the Stars— Focus on Earth's Rotation

by Richard Russo

Star trails tell tales about the movement of our planet.

People have always been aware of the pattern of night and day, of darkness and light. Early civilizations relied heavily on this cycle to keep track of time—counting, for example, the number of "suns" that a journey took, or the number of "suns" before the big hunt. Some of these peoples spoke about the Sun's moving across the sky, perhaps driven by a chariot.

We know now, of course, that it is the eastward rotation of the Earth that makes the Sun appear to be moving westward across the sky. And we can use the Sun, and other stars, as tools to observe that rotation with some precision.

The apparent daily movement of the Sun around the Earth is called *diurnal motion.* It is because the Sun is so close to us that its movements are detectable. But in fact, all stars, all planets—in other words, all celestial bodies—describe diurnal circles around the Earth.

One way of studying the Earth's rotation is to track the diurnal motion of the Sun. The brightness of the Sun makes it dangerous to have students measure its movements directly. A sundial provides a safe way to track

the Sun's movements, although it does so indirectly. Sundials are familiar to most students and are easy to use, but they cannot match the accuracy that we gain by using another relatively simple technique—astrophotography.

Your students don't need an extensive astronomy background or fancy telescopes to do this activity. All they need is a 35-mm camera loaded with 200 ASA print film, seated on a tripod, and aimed at the celestial sphere above.

As "terra firma" rotates, the positions of the stars relative to the Earth change. The camera records the path of the stars in the sky above as streaks, or trails, on the film.

Center your "astro-camera" on the star Polaris. This view will provide you with some very informative star trails. Why is Polaris so important? If we were to extend the north ray of the Earth's rotational axis about 200 light-years into the Milky Way, we would find that Polaris lies very near this extension. In other words, within 1 degree, Polaris is fixed at true north. This gives the star tremendous navigational significance as a stable point of reference in the celestial sphere.

For us, Polaris, or the North Star, is important because the surrounding

Richard Russo is the science coordinator for Montvale Schools, Montvale, NJ 07645.

Earth at Hand

137

stars appear to revolve around it as they describe their diurnal circles. Again, this phenomenon stems directly from Polaris's position at the end of the Earth's axis of rotation.

We can take advantage of this phenomenon to observe the Earth's rotation. All of us know that circles are divided into 360 degrees. Earth takes one solar day, or 24 hours, to describe the full 360 degrees, or to rotate once around its axis. If we were to leave the shutter of the camera we have centered on Polaris open for 24 hours, we would see a dim dot surrounded by many concentric circles. These circles are the trails left by the stars seeming to revolve around Polaris, the dim dot. Naturally, if we left the shutter open for a shorter period of time, our picture would only reveal arcs, or shorter segments, of those circles. (See photo.)

To calculate the Earth's period of rotation, all we have to do is open the shutter for a set period of time and measure the arcs in the resulting picture. The number of degrees in a circle divided by the number of degrees in the arc multiplied by the time elapsed during the picture taking gives the period of the Earth's rotation.

Let's get to the activity. Preparations and calculations can be made during class, but the camera work will have to be done at home at night. This gives the project a strong parental component. Because of the logistical complications of scheduling and monitoring evening sessions, clear specific directions will have to supplement your assignment.

In addition to a camera, film, and a tripod, each student will need a cable release, a small flashlight, and a stopwatch. Light is this activity's greatest enemy. About 90 minutes is the limit for exposures taken within 50 kilometers of a metropolitan area, because artificial lights "pollute" the night sky, making the stars harder to see. Therefore, students must choose a moonless night and a location as far as possible from any light sources.

Students simply set the focus of the camera at infinity and the time exposure at "b" for indefinitely long exposures. They insert the cable release, and sight the North Star into the mid-

dle of the camera's field of view. Suggest they use the "pointer" stars of the Big Dipper's handle to locate Polaris. Once Polaris is in the field, they depress the cable release and lock it. The shutter is now open for whatever length of time you require.

For the first part of the activity, have your students take a photograph with an exposure time of 1 hour. After their film is printed, they should place a protractor over the photograph (with the center reference point of the protractor on Polaris) and measure the angle covered by one of the star trails. Be sure to point out that the angles of the long trails made by stars distant

—*Photo courtesy of the author*

from Polaris are equal to the angles of the smaller trails made by stars closer to Polaris. The closer stars simply form "stubbier" triangles.

Using the line of reasoning discussed above, your students can calculate the Earth's period of rotation from their print. They should determine that, since stars "move" approximately 15 degrees in 1 hour, the period of the Earth's rotation is approximately 24 hours.

Now have your students do some calculations to show that 1 degree of rotation is equal to 4 minutes of time. (It takes Earth 24 hours to rotate 360 degrees. If the students divide 360 by 24, they will get 15 degrees per hour, or 15 degrees per 60 minutes. A star that covers 15 degrees in 60 minutes will cover 1 degree in 4 minutes.)

To complete this activity, again ask your students to take photos of the

inimitable North Star and the surrounding north celestial polar region. The constellations around Polaris are called *circumpolar*. These constellations do not set for most observers in the Northern Hemisphere. A rather interesting formula is used to determine which stars are circumpolar for your latitude.

If a star is nearer the celestial pole (that is, true celestial north) than the pole itself is to the horizon, the star does not dip below the horizon; hence, it is circumpolar. Consequently, for an observer in the Northern Hemisphere, a star never sets if its north polar distance (90 degrees minus the star's latitude) is less than the observer's latitude.

Here's an illustration: The bowl of the Big Dipper is at celestial latitude 58°N. The Big Dipper never sets at terrestrial latitude 40°N (New York City) because its north polar distance of 32 degrees is less than the latitude of the Big Apple.

You should have students outline the circumpolar constellations and highlight their trails in the same photo. This technique is a bit tricky! Have the students aim at Polaris as before and open the camera's shutter for 10 seconds only. Without closing the shutter, have them cover the lens with a hat for 2 minutes, and then remove the hat for 20 minutes. When the film is developed, their prints will show defined stars, followed by spaces of 0.5 degree (about the diameter of the Sun and the Moon), followed by 20-minute, or 5-degree, trails. (The trails are the same length as the distance between Mirak and Dubhe, the pointer stars of the handle of the Big Dipper.)

If you want your students to appreciate fully the swiftness of our planet's rotation, just tell them to watch the setting Sun or Moon. Remember that the Earth moves about 1 degree in 4 minutes of time. The Sun and Moon each have apparent angular diameters of 0.5 degree. The next time students observe the disk of the Sun or the Moon just making contact with the horizon, they should notice how quickly it sets. How quickly? Only about 2 minutes! That's why the Sun is always gone when we run to get our cameras to catch those beautiful sunsets. ∎

Build a Portable Planetarium

You can build a portable planetarium that will accommodate 25 students by using large pieces of black plastic to form an inflatable bag. From the interior of the bag, the class can visualize stars and constellations by light filtering through holes punched in the bag.

To begin, hang a large piece of black garden plastic (about 5 meters by 5 meters) on the wall. With an overhead projector, project a transparency traced from a star chart onto the plastic. One source for the star chart is "The Evening Skies" (courtesy of Abrams Planetarium, Michigan State University) which appears in NSTA's elementary journal *Science & Children*. The darkness of the plastic may require may require one person to point to the stars and a second to affix small pieces of masking tape to the plastic.

After marking the locations of the stars, removed the plastic from the wall and punch holes where the tape is. A punch that can make holes of variable diameter will let you represent the relative brightness of the stars. Pinch the plastic, slip the punch over the fold, and cut a half moon in the plastic. Release the plastic, and you have a circular hole.

Next, cut a second piece of plastic (about 4 meters by 3 meters), and lay the two pieces of plastic on top of each other. Put tucks in the larger piece so that the corners of the two pieces match, and then tape the pieces together on three sides, forming a bag. Turn the bag inside out, and tape the exterior of the three seams with black electrical tape or duct tape for reinforcement.

Part of the fourth side is the doorway to the planetarium. Tape the edges of the fourth side, closing only the area near what will be the top of the planetarium later. Overlap the edges of the doorway so that you will be able to trap air in the bag. In the side opposite the doorway, cut a small slit to allow entry of the filler tube, which you will use to inflate the planetarium. Tape at the bottom and top of the slit will reinforce it.

Cut a third piece of plastic (about 2 meters by 2 meters), and roll and tape this piece to form the filler tube. To inflate the planetarium, insert one end of the filler tube into the bag. Attach the other end of the filler tube to a fan using velcro pieces that have been glued with epoxy to the fan and the filler tube. The diameter of the filler tube should match the circumference of the face of the fan.

Because the planetarium's materials can be costly, you probably will have to sell parents, the school, or contributors on the idea. To do this, you can use a garbage bag to build a scale model that you can inflate with a hairdryer.

Libby Lawrie
Ben Davis Jr. High School
Indianapolis, IN

Proving that the Earth is Spherical

Pythagoras, the Greek philosopher, is credited with being the first person to believe that the Earth was a sphere. One way to prove that this is true is a simulation of a ship sinking over the horizon.

Prepare a cardboard sailboat as shown in Figure 1. Then, tape one end of a large piece of poster board to a table top. Secure the untaped edge of the poster board about 15 centimeters above the table so that the paper curves, as in figure 2. Have students position themselves so that their eyes are just at the height of the raised edge of the paper. Move the cardboard sailboat over the paper, and have students record their observations. In clear weather, the students can observe the same phenomenon at the horizon of an ocean or large lake.

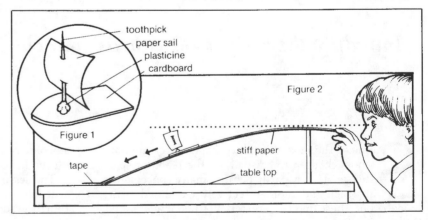

Art by Max Karl Winkler

Ovid K. Wong
Elk Grove Village, IL

Playing with Planets

Introduce the concept of scale as your class explores the solar system.

Some of the most difficult concepts for elementary school children to grasp are in the realm of space science. This is partly because many concepts are abstract and more effectively explained using models, but these models are rarely student generated. Space science becomes merely show-and-tell, with the teacher playing the leading role.

The following lessons bridge the gap between concrete and abstract by introducing students to the valuable idea of *scale*. Your class will learn about the relative size, position, and orbital patterns of the planets in our solar system using mathematics to construct models of objects that they cannot see. Many upper elementary and middle school students still require a concrete representation of an abstract concept; it even helps adults sometimes.

The lessons proposed here can be "scaled" to fit your schedule and your students' ability levels. For instance, a fourth grader might take longer to apply the concept of diameter measurement than would a fifth-grade student. Allow enough time for students to discover and construct without being rushed.

For Fourth Grade

For several days before you actually begin this lesson series, discuss with your class diameter measurement and the metric system. Prepare work sheets on which students

can practice measuring and specifying the diameters of different-size circles.

Ask students for examples of scaling. Guide them in a discussion about why we must use a scale when working with the solar system. Identify the scale you will be using as 1 cm = 10,000 km. It might be a good idea to take the class outside and measure off 1 km. Or tell them that 1 km is about the length of five city blocks. Point out that 10,000 km is about twice the coast-to-coast distance in the United States.

This activity on identifying the planets' diameters requires the following materials for each student:

- a metric ruler,
- crayons,
- scissors,
- glue,
- a chart listing the planets' scaled diameters (see fig. 1),
- and a work sheet with circles of specified diameters and spaces alongside each circle.

Distribute the diameter charts and the work sheets. Tell students that they can complete their work sheets, identifying the circles by planet name, using information from the chart and their rulers.

A follow-up discussion could include such questions as: Which planet is nearly the same size as Earth? Which planet is the smallest? The largest? How many planets are larger than Earth? How many are smaller than Earth?

By Melanie Mann-Lewis

Next, have the students color their planets, cut them out, and paste them on a piece of paper, from left to right in order from the sun. In this lesson, we are concerned with order, not distance. Students might want to make mobiles of the solar system using string and coat hangers. If you like this idea, draw the circles on heavy paper.

On the playground or any asphalt-covered area, students can use a string and chalk to draw the sun using the scale 1 cm = 10,000 km. Measure a little more than 695 cm (6.95 m) of string to represent the radius of the sun. Before cutting the string, tie one end to a piece of chalk and the other end to a student's finger or to a pencil. The scaled measurement should be accurate *after* tying the string. Instruct one child to hold his finger or the pencil attached to the string stationary while another child walks around him, tracing the circumference of the circle on the asphalt. Keep the string taut. Allow students to compare the drawings of the planets from the previous day's assignment with the drawing of the sun on the playground. Let students hypothesize about the diameter of the sun they've drawn. They might form a generalization about the relationship between radius and diameter.

Fifth Grade and Up

With only string, pencils, compasses, and cardboard, students can use the radius of a planet to construct a spherical representation.

A different scale will be required for students to draw the small planets accurately. In the following activity, 1 cm = 1,000 km.

For a few days, have the class review the concepts of radius, diameter, and circumference. Work sheets requiring students to use the diameter of a circle to determine its radius and circumference will be most helpful in preparing them to apply these concepts in the following activities.

Have students discuss how the knowledge of planet diameter benefits scientists involved in space exploration.

Each student will need

- a metric ruler;
- scissors;
- a compass;
- string or yarn;
- construction paper, butcher paper, or aluminum foil;
- colored markers;
- and a work sheet listing the planets' scaled diameters with spaces for radii and circumferences (see fig. 2).

Distribute the work sheets. Ask students to calculate the radius and circumference of each planet and to fill in the appropriate spaces on their work sheets.

Then, instruct students to work in groups of two or three. Each group must draw and cut out two circles of one planet's scaled diameter. Distribute string, compasses, and cardboard. Students should set their compasses or measure their strings (if constructing the larger planets) to the length of the radii they calculated on their work sheets. This reinforces the concept that diameter is twice the radius.

For the inner planets and Pluto, students can use a compass. (Some safety precautions might be necessary before working with the compasses.) If you have access to an oversized wooden chalkboard compass, use it to construct Uranus and Neptune. For Jupiter and Saturn, use string

---

Figure 1

1 cm = 10,000 km

Body	Diameter in cm
Mercury	0.4
Venus	1.2
Earth	1.3
Mars	0.7
Jupiter	14
Saturn	12
Uranus	5
Neptune	4.9
Pluto	0.5*
Sun	1,390

*Because Pluto is so far away, it is difficult to calculate its size accurately. Some texts will say that Mercury is smaller than Pluto, while other texts indicate the opposite. Discuss this with students.

Figure 2

1 cm = 1,000 km

Body	Diameter in cm
Mercury	4
Venus	12
Earth	13
Mars	7
Jupiter	140
Saturn	120
Uranus	50
Neptune	49
Pluto	5
Sun	13,900

---

and pencil in a manner like that used to draw the sun in the lesson for fourth graders.

After each group has drawn and cut out its two circles, the students can put the two circles together for a three-dimensional effect (see fig. 3). Planets should be labeled with their names and diameters. Allow students to cover

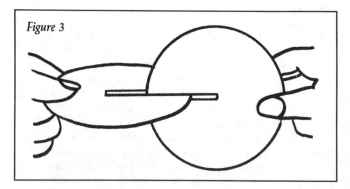

Figure 3

> Challenge students to devise
> a scale for orbital time
> that they could match
> in their dynamic model
> of the solar system.

their planets with colorful construction paper, butcher paper, or aluminum foil. You may want to hang the planets from the ceiling in the proper order from the sun.

Using the string-and-chalk method, students can draw the sun on an asphalt surface outside. Students may use yellow butcher paper to trace a slice of the radius, cut it out, and bring it back into the classroom.

Students will be amazed when they compare the radius of the sun to the diameter of the nine planets. Have students calculate how many times larger the sun's diameter is by dividing the diameter of the sun by that of each planet. Some students may be led to discover the area of the circle representing each planet and the sun (pi × radius²). Give students the formula 4 × pi(radius)² and let them calculate the approximate surface area of each sphere.

Now, ask the class several discussion questions:
- The greater the mass of an object, the greater the gravitational force it produces. If the planets' density were constant, which planet would have the greatest surface gravity? (Jupiter.)
- How does gravitational pull affect your weight on the surface of a planet? (Weight is a measure of gravity pulling on an object. Therefore, if density were constant, on the surface of bigger planets you would weigh more.)
- Would you weigh more on the surface of Mercury or Earth? (Earth.)

Next, let your class become a dynamic model of the solar system. Select students to represent the sun and the nine planets. They can identify themselves with name cards or the model planets they made earlier. The scale for this activity will be 1 cm = 8,000,000 km (see fig. 4). You'll have to take the class outside for this activity.

Let the students measure off the appropriate amount of string for their planet's distance from the sun. Make sure students wind the string onto cylinders (paper-towel tubes or pencils work well) to avoid a tangled mess.

The student who has been selected to be the sun will hold a sturdy pole or wooden meter stick labeled "Sun." The "planets" should tie their strings securely to the pole. One at a time, each planet, starting with Mercury, should pace off its distance so that other students can relate a distance with a particular planet.

After all of the planets are in place, signal them to begin orbiting the sun. To avoid tying up the student labeled

Figure 4

1 cm = 8,000,000 km

Planet	Distance from Sun in cm
Mercury	7.2
Venus	13.5
Earth	18.7
Mars	28.5
Jupiter	97.3
Saturn	178.1
Uranus	358.4
Neptune	560.8
Pluto	736.3

Figure 5

D = days, Y = years

Planet	Sidereal Period
Mercury	88 D
Venus	224.7 D
Earth	365.3 D
Mars	687 D
Jupiter	11.9 Y
Saturn	29.5 Y
Uranus	84 Y
Neptune	164.8 Y
Pluto	247.7 Y

"Sun," have her raise the pole high above her head.

After the planets have gone out of orbit, ask some questions. What is the difference between a revolution and a rotation? If distance from the sun were the only factor affecting temperature, which planet would be hottest? Coldest? If all the planets traveled around the sun at the same speed, which would have the longest year? The shortest?

When you return to the classroom, students might be interested in seeing the sidereal period chart (see fig. 5), which measures time and motion relative to the fixed stars and gives the time period necessary for one complete revolution around the sun. You may want to challenge students to devise a scale for orbital time that they could match in their dynamic model of the solar system.

These activities should help you provide students with information on the solar system, practice in applying mathematical concepts to solve problems, and a working knowledge of scale.

Melanie Mann-Lewis, formerly science curriculum specialist for the Memphis (Tennessee) City Schools, is assistant principal at Alton Elementary School in Memphis. Artwork by Tricia Tusa.

MAKE A MOON CALENDAR

"Hey, I saw a gibbous moon while I was waiting for my school bus today!

"The crescent moon was out when we were running laps for P.E."

"I saw the full moon coming up last night!"

If you think these remarks sound too enthusiastic to be coming from typical middle school earth science students, try making a moon calendar with your class. This easy, hands-on activity introduces students to a unit about the moon, sparks their interest, and produces a handy reference tool usable throughout a moon unit. In addition, students gain valuable practice using sequencing and spatial relationship skills.

A finished moon calendar has 30 days, begins and ends on a day with a full moon, and has a picture of each day's moon phase glued onto a calendar square. Students study the calendar to discover when the moon is waxing or waning, crescent or gibbous, or when a quarter, full, or new moon is present.

Preliminaries

To prepare your handouts, first consult an almanac or calendar to find out when the current month's full moon occurs. Copy that month's calendar page plus the days from the previous or following month needed to show one complete moon cycle. Piece together your calendar, making sure it includes two full moons and the actual dates of your unit.

Next, make copies of the 30 moon pictures shown on the next page. For this activity, the 29 1/3 day cycle is represented by 29 moon pictures plus one additional full moon picture. Make sure that you size the moon pictures to fit inside the daily calendar squares without covering the dates.

The Activity

Allow one class period to complete the activity. Each student will need a prepared calendar, a moon picture page, scissors, and glue. The students' task is to cut apart the 30 moon picture squares, sequence the pictures correctly, and then glue each one onto the correct calendar day. Tell the students that when the moon pictures are in the correct order, the letters in the centers of the squares will form the sentence: "The moon revolves around the Earth!"

Stress that the shape and size of each day's moon must fall midway between the previous and the following day's moons. Trouble-shoot by pointing out the different moon shapes on the five letter Es, and tell the students that they will need to sequence the shapes carefully. Write the *dates* for both full moons and the new moon on the board to give three reference points.

Allow thirty minutes for cutting and sequencing and an additional ten minutes for gluing. It should be easy for you to scan for accuracy by checking to see that the moon shapes get progressively smaller from the first full moon to the new moon and progressively larger from the new moon to the second full moon. To guarantee student success, hand out the glue only after you have checked for correct sequencing.

Uses and Applications: Shoot for the Moon

Have students keep moon observation charts at home. They can record observed moonrise and moonset times and make sketches of the moon's profile and visible topographic features. (Encourage the use of binoculars.) Ask students to record the compass direction on the horizon where the moon rises and sets. After several days, students can pool their observed data in class, compare them with the moon calendar predictions, and attempt to reach some conclusions. For example, they can conclude that the moon rises in the east and sets in the

Moon Calendar Questionnaire

Directions for Students: Use your moon calendar and the following definitions to answer the questions below.

- Quarter Moon: When exactly half of the full moon is visible.
- Crescent Moon: When less than one quarter of the moon is visible.
- Gibbous Moon: When more than one quarter of the moon is visible.
- Waxing Moon: When the moon is increasing in intensity and size.
- Waning Moon: When the moon is diminishing.

1. Look at the shape of the moon on the six days following the first full moon.
   a. Is it a crescent or gibbous moon?
   b. How many days after the first full moon does a quarter moon appear?
2. Write the dates when a crescent moon is visible. (Find two different sets of dates.)

3. Write the dates when a waxing moon is visible. (Find two different sets of dates.)

4. Write the dates when the moon is:
   a. Waxing Gibbous:
   b. Waning Gibbous:
   c. Waxing Crescent:
   d: Waning Crescent:

5. Why does the new moon appear to be invisible to people on Earth?

6. If you count the first full moon as Day 0, how many days are there until the day of the new moon? How many days are there from the first full moon until the second full moon?

7. The phrase "Once in a blue moon" describes an event that happens very rarely. It comes from the term "blue moon," which describes a month that has two full moons. Is this a blue moon month? Why or why not?

National Science Teachers Association

Moon Pictures

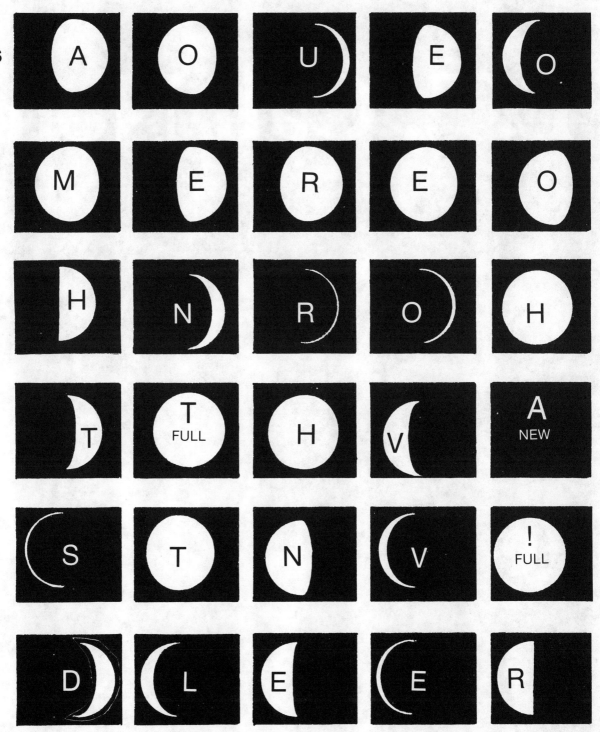

west. This is the scientific method in action!

Have your students turn to the calendar to find real-life examples for the moon phase concept covered in the textbook. Then use questions based on the calendar to check for understanding. (See Moon Calendar Questionnaire.)

Consult your newspaper or almanac to get one day's moonrise and moonset times. Using these times, have the students calculate moonrise and moonset for the rest of the week or month. Adding 50 minutes to the times for day one gives

the times for day two, and so on. Ask your students to record this data on their calendars, and encourage them to watch the horizon at these times.

Use a local tide table to find the two high tide levels for each day of your calendar month. Graph tide levels for each day of the month, and then ask the students to determine the relationship between moon phases and tide levels.

The moon calendar helps to make a science concept part of a student's daily experience. It is a lesson that integrates

basic reading skills with science concept learning. It allows students to practice observing patterns and making predictions about the moon's monthly display. You'll find yourself using this activity much more than "once in a blue moon!"

SUSAN BALLINGER
California State University, Fresno
Fresno, California

E.T., Call Nome

. . . or Phoenix, or Burbank

Let's just hope they don't call collect.

With all of the radio and television signals "leaking" from Earth into outer space, is it possible that someone or something has received them? Perhaps at this moment, they are trying to locate us. If so, do we have the technology to understand what they might be sending us?

For years, astronomers have been listening for a reply. They've been monitoring certain stars for radio-wave messages, as well as beaming radio waves back to them. Since 1960, scientists have monitored Tau Ceti and Epsilon Eridani—two stars that are less than 12 light-years from Earth—for radio messages. But, no luck so far.

In Puerto Rico, astronomers have used a large radio telescope as a transmitter throughout the galaxy, sending a steady message from Earth. They've been broadcasting the atomic numbers of our basic elements, the binary code of DNA, the average size of an adult human being, and a description of our solar system to star cluster M–13 in the constellation Hercules. Presumably a civilization advanced enough to receive the messages would be advanced enough to understand them. Again, we have heard nothing, but that may have something to do with distances.

M–13 is 21,000 light-years from Earth. What that means is that a radio-wave message broadcast in 1987 and traveling through the universe at the speed of light—186,000 miles per second or six trillion miles per year—will be received at the star cluster in 22,987 A.D. Assuming someone would be there to receive the message—someone who can then translate it into fluent M–13ish—and assuming that someone would want to answer immediately, we wouldn't get an answer until 43,987 A.D. And that might explain why we haven't heard from anyone yet. The mails take time.

The size of the universe can be dumbfounding. If we consider our own solar system to be the size of, say, a coffee cup rim, the Milky Way would be the size of the entire North American continent. Conceptualizing the universe, with its "billions and billions" of stars and solar systems—some of them probably inhabited—can overwhelm a young student, just as it does an adult. But the prospect of exchanging messages with extraterrestrial beings—no matter how many light-years away—is especially exciting for young students.

To help students understand their relationship to and their place in the universe—as well as how difficult it would be to find life beyond Earth—you may want to try some of these classroom ideas. They are all meant to be thought-provoking about Earth's uniqueness in our own solar system.

Map Procedure

- Use a city street map and have students locate their homes on it.
- Use a state highway map and have the students find their town on it. Have them locate other towns on the map and then give directions to these places.
- Using a poster-size map of the United States, have students locate their state and town on the map.
- Place a world map next to the United States map, and have the class locate the United States, their state, and their town. Have the students imagine they're visitors from outer space and that they are approaching Earth in a spaceship. Show your students a color photograph of the planet Earth. What is the main color? Does our planet look as if it's essentially a water world? Where do you think an extraterrestrial would look for intelligent life on Earth? Why?
- Using a poster of the solar system, locate the Earth. Then, if possible, locate the North American continent and the approximate location of your state.
- Use a picture or drawing of a spiral galaxy to illustrate what astronomers think is the shape of our galaxy, the Milky Way. We live in the spiral arm of the galaxy, in a region called the Orion Arm. Imagine the difficulty of describing what our galaxy looks like from within it. It is like describing the outside of a house while sitting in the

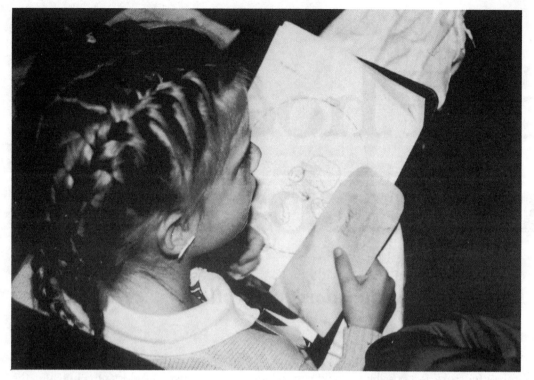

living room. Our understanding of the spiral nature of our galaxy comes from studying other galaxies and using them as models for comparison.

• To take this one step further, take several pictures of differently shaped galaxies, and place them together as a cluster of galaxies around the Milky Way.

The Model Procedure

You can use models instead of maps. Substitute a globe for the map of the Earth. Use a solar system model to illustrate Earth's place in the solar system. Models of the planets can be the result of an art project. Have students decorate Styrofoam balls to represent the planets. Use cotton balls and pipe cleaners to make models of the galaxies.

While working with your students on this activity, remember that the models will not be the correct size, relative to each other. This could be corrected, at least for our solar system, by working out scale sizes of the planets, as well as their relative place in the our solar system [see boxes, below].

Planet	Distance in Model if Sun Is Size of Thumbnail
Mercury	1 baby step (three baby steps=one giant step)
Venus	2 baby steps
Earth	1 giant step (one giant step=one meter)
Mars	1 ½ giant steps
Jupiter	5 giant steps
Saturn	10 giant steps
Uranus	20 giant steps
Neptune	30 giant steps
Pluto	40 giant steps

Using the scale at right, the Sun would be the size of a thumbnail and Jupiter and Saturn would be the size of a small freckle. The other planets would be the size of a hair's diameter or smaller.

Planet or Sun	Diameter of Model In Units
Sun	285
Mercury	1
Venus	2½
Earth	2⅞
Mars	1⅜
Jupiter	29
Saturn	24½
Uranus	10½
Neptune	8¼
Pluto	½

Planet or Sun	Diameter in km	Distance from Sun in km
Sun	1,392,000	
Mercury	4,878	57,900,000
Venus	12,104	108,100,000
Earth	12,756	149,500,000
Mars	6,794	227,800,000
Jupiter	142,796	778,000,000
Saturn	120,000	1,427,000,000
Uranus	50,800	2,869,000,000
Neptune	48,600	4,497,000,000
Pluto	3,000	5,900,000,000

Start your classroom model with the Sun in one corner of the room. As you place the planets in their scale distances from the Sun, you might find yourself outside the classroom moving farther and farther away, depending on the scale you chose. If your scale goes well beyond the school's boundaries, determine the orbits of the outer planets by noting the streets around the school that the orbits would cut across. Have the students note these orbital positions as they travel to and from school.

Good Ol' Sol

Sol, or the Sun, is our own star. It's an average-sized, yellow star, and unlike other stars in the skies, which usually occur in groups of two or more, our star is a loner. Perhaps this is a prerequisite for life to develop on a planet. Yet, Sol may have had an opportunity for a companion star when the solar system was first forming.

Jupiter is like a star in some ways: It's a large ball of gases and a source of radiation. But, unlike active stars, Jupiter doesn't have nuclear reactions occurring at its core.

What would it be like to have two stars in the sky? How would shadows look during the day as the Earth rotates? How about seasons? Would they be the same with two stars? Would we have double sunrises and sunsets?

Activities such as these do more than provide students with facts and figures. They make students use their hands and eyes and their minds.

We are made of the stuff of stars, someone once said, and it is our destiny to one day travel among them. We pave the way there by fostering in our students an interest in space.

Resources

Allison, Linda. (1975). *The reasons for seasons.* Covelo, CA: The Yolla Bolly Press.

Burns, M. (1978). *This book is about time.* Covelo, CA: The Yolla Bolly Press.

Chisholm, Jane. (1982) *Finding out about our Earth.* Tulsa, OK: EDC Publishing.

Myring, Lynn, and Snowden, Sheila. (1982). *Finding out about the Sun, Moon, and planets.* Tulsa, OK: EDC Publishing.

"Astronomy Adventures." Ranger Rick's NatureScope. Washington, DC: National Wildlife Federation.

Ryder, Joanne. (1984). *Earth and space.* Racine, WI: Western Publishing.

Bob Riddle, a former science teacher, is currently director of Project STARWALK, an Earth-space science program for students in grades 3–6. Available for schools nationwide, Project STARWALK combines classroom and planetarium lessons to help students develop skills in observing, graphing, and predicting. Photographs courtesy of the author.

Space Race

JAMES M. HEROLD

Send students off on a challenging, exciting mission that requires knowledge of space and imagination.

Imaginations soar when students design and build model spacecraft for a space race. Fantasy and a unique brand of practicality merge to create fantastic ships with diamond windows or spider-web solar panels. Huge springs power one craft, while another sheds its skin with each new environment. Each student group creates its own bizarre but practical solution for a space race.

It can be a mind-boggling challenge for students to consider all the consequences and challenges associated with space exploration. Your students cannot solve the difficulties of space flight realistically, but they can begin to appreciate the multitude of preflight decisions made for every space journey in a fun and sometimes bizarre space race.

Groups of students design and build models of their own spacecrafts for a race to five space bodies: Earth, the Moon, Io and Callisto (two of Jupiter's moons), and Mars. Using common sense solutions, students confront and resolve the same problems as aerospace engineers. Their projects may not be technical masterpieces, but students will have solved the problems of space flight in a unique way. Through research activities meeting their own needs, students gain more knowledge of our solar system and space flight than lecture could possibly provide. To successfully compete, students' space craft plans must address the

- physical challenges to the safety of the vehicle and the crew;
- schedule of stops and reasons for this particular itinerary;
- energy source the vehicle will use,
- crew size and how size is determined;
- adaptability of the craft as it functions in various environments; and
- visual appearance of the craft fulfilling both form and function.

The possibilities

Students explore space as they try to solve the practical problems of space travel for themselves. You can show them realistic possibilities for the design of long-term, manned spacecrafts.

For instance, the available energy sources include rocket power, nuclear energy, solar power, gravity assist, and solar wind. At present, students often select rocket, solar, and nuclear power. Each source has many advantages and disadvantages, depending upon the circumstances. Solar power offers fewer advantages as the craft

---

James M. Herold is a science and art teacher at Spring Valley High School in Spring Valley, Wisconsin.

moves further from the Sun. Rocket power poses weight problems as the number of launches increases.

For interplanetary travel, gravity assist is effective, as seen in Pioneer 10 and 11 and Explorer 1 and 2. The craft uses the gravitational pull of a planet or its satellites, while avoiding total gravitational entrapment, to increase its momentum. As inertia and centrifugal force increase, the craft slingshots forward at a faster speed.

Another source of energy is solar wind, a rather avant-garde possibility, suggesting that a ship can set up a sail to harness the solar wind.

Students must also carefully consider crew size for this voyage. No one can predict what will happen during a long-term flight. Illnesses may strike, accidents can happen, and boredom and listlessness can develop. Some physical conditions might even result directly from long periods in space, such as muscle degeneration resulting from zero gravity.

The craft itself must be both light and durable. It should be light enough for several take-offs against various gravitational fields; yet, it must also be able to withstand the stresses of temperature extremes and other possible weather conditions.

Teacher preparation

1. Begin by gathering research sources for students since they will have many questions to answer before they can successfully design their crafts. Allow students to find the necessary information within sources or possibly within their own textbooks.

2. Explain the rules of the race to the groups. They are to design and build a model of a manned spacecraft for a voyage to five space bodies: Earth, the Moon, Io, Callisto, and Mars, in whatever order they choose.

3. Prepare to evaluate this activity as interdisciplinary work. Consider English skills in oral and written reports, art skills in the model, and of course, science skills in evaluating the entire project. The project has the capacity to emphasize art and creative skills, thus it ecnourages students to perceive the creative aspects of science. See the Project Worksheet for suggestions, or solicit help from teachers in other disciplines.

For the science evaluation, the Project Worksheet lists some criteria that each project should meet. Consider how well the models fulfill these requirements and what problems there might be with those solutions. When problems occur, decide if they are at the students' level. If not, do not let it influence the group's grade.

Introduction

A space race is underway in the galaxy, and your space crew has a mission to design the ultimate space machine. It will be the fastest, safest ship in space as it maneuvers to pit stops on Earth, the Moon, Io, Callisto, and Mars. The design of the ship, order of the pit stops, source of power, and crew size and individual assignments are up to your own creative genius. The only restrictions are that all five pit stops must be made, the ship must be manned, each crew member must be on board at each stop, and ships cannot interfere with the progress of other competing ships.

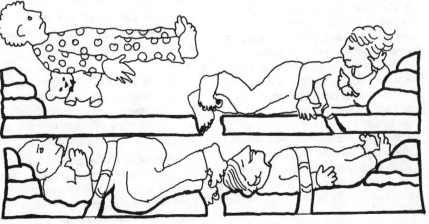

Art by Alice Faith Dole

Table 1.

Body in Space	Diameter Km	Surface Pressure atm	Density g/cm³	Mass x 10²⁴ Kg	Surface Gravity cm/sec²	Surface Temp. °C
Earth	12,756	1.0	5.52	5.98	981.0	25
Moon	3,476	0	3.34	0.74	1.57	-130
Mars	6,786	.0007	3.39	0.64	373.0	-60
Io	3,636	10^{-8}	3.55	0.089	177.0	-148
Callisto	5,820	0	1.82	0.106	118.0	-130

Procedure

1. Before beginning your design, use resource materials to consider the nature of space.
• How do objects or particles usually travel in space?
• Identify methods by which an object might be influenced to move in space.

2. Look at Table 1, it compares environmental conditions for the five space bodies. As a result of these conditions, there are hazards and advantages. List the hazards and advantages for each space body. Consider how you can utilize the advantages and avoid the hazards.

3. The rules require that each craft be manned. Consider how many crew members to take on the voyage and what effect each crew member's presence would have on your design.
• What tasks or assignments are necessary for each crew member to perform while this race is in progress?
• How many crew members will be needed to perform these tasks?
• What necessities must be provided for the crew to remain physically and psychologically healthy?

4. Consider how you can make sure that your craft survives the stresses of temperature, weather, and speed during this rigorous journey.
• What materials are durable? Flexible? Heat resistant?

5. With the previous information in mind, begin designing your spacecraft.

6. Decide your route. Research the five space bodies. What factors will influence your route?
• Which route is the shortest?
• Which stops will be the most difficult?

7. In a written report, explain your design choices and race route.

8. Prepare a model of your spacecraft. Make it as aesthetically pleasing as possible without sacrificing function.

9. Prepare and present an oral report on your spacecraft and race route. Tell the class why your craft should win the race. Involve each member of the group in the presentation.

10. Prepare to challenge any faulty design features in the models of other groups. If the teacher agrees, the craft is grounded, giving the other teams one less opponent.

Teacher evaluation

You must decide if the craft should be grounded. When all models have been presented, decide the winner of the race. (Use the Project Worksheet to help you decide which group best addresses all of the problems of long-term space travel—this group should win the race.)

For each group, or for the class as a whole, you might write a creative piece of your own, telling students what happened when their craft was used. Phrase your report from a historical perspective and throw in a little humor, and students will really enjoy the bizarre ways that they might win or lose.

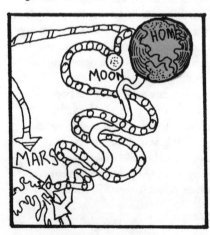

Project evaluation sheet

Science Content

_____ Power sources used in this ship are explained so that the reason for use and how they work is understood.

_____ Crew assignments and size are explained so that it is evident some consideration was given.

_____ Environmental hazards of each pit stop are correctly identified and associated with each pit stop.

_____ Itinerary of pit stops shows an understanding of the fastest route to travel and location of pit stops in planetary motion.

_____ Sales pitch on the ship is convincing in how the ship might win this race.

Mechanics

_____ Spelling

_____ Punctuation

_____ Sentence structure

_____ Content

Presentation Design

_____ Format is well used.
Illustrations should fill the page, even touch the edge of the paper, rather than hiding the details of the illustrated subject in a tiny drawing.

_____ Emphasis is easy to see with the center of emphasis.
The eye needs a natural area to rest. By controlling shading and color, one area in the drawing stands out as an "emphasis."

Architectural Design

_____ Interior space shows consideration for human needs.

_____ Exterior space reflects concern for visual appeal through interesting arrangement of form.

Technique

_____ Presentation ready
Presentation appears complete and ready for public display.

_____ Craftsmanship
Concern for neatness and style is apparent.

Creativity

_____ Design is original.

_____ Design is creative.

_____ Design appears feasible.

Earth at Hand

MAKING WAVES

Ormiston H. Walker

The next time you're stopped at a railroad crossing, waiting for a freight train to pass by, think about Christian Johann Doppler. The nineteenth-century Austrian scientist is not responsible for elucidating the mechanics of motion that propels the train forward, but he did explain why it is that sound waves (in your case, the ones emitted by the whistle on the approaching train) seem to change frequency when the source of the sound is in motion and the listener is stationary. The phenomenon, called the Doppler effect, accounts for the fact that you hear a shrill, piercing sound as the train approaches and a low mournful one as it recedes in the distance.

Don't You Hear That Whistle Blow?

The scientific explanation of this phenomenon is straightforward enough. Suppose a freight train traveling at about 70 kilometers per hour sounds its whistle at 440 hertz, a frequency that corresponds to A natural on the musical scale. The first wave reaches you as you sit waiting in the car. But in the time it takes for the second wave to arrive, the train has moved forward,

Ormiston H. Walker lives in Christchurch, New Zealand. Diagrams by Johanna Vogelsang.

and the second wave is closer to the first than it would have been if the train had been standing still. The third wave is closer yet, and the fourth, and the fifth. As long as the train continues to move toward you, the waves crowd in on one another. So instead of receiving 440 waves per second (as you would if both you and the train were stationary), you receive 466 waves, a frequency that corresponds to an A sharp, a half tone higher than the original A natural note the engine sounded.

As the engine passes, you'll hear the sound of A natural, but only for a fraction of a second. Then as the engine recedes, the sound waves begin to lag behind and lengthen. The second wave is further from the first and the third is further still. The frequency of the waves decreases, so the sound you hear is lower in pitch than the sound you heard as the train approached and as it passed by. Instead of 440 or 466 waves a second, you hear only 415, a sound that corresponds to A flat. This change in the frequency of the sound that reaches your ears is called the Doppler effect.

Waves of Starlight

The Doppler principle is used to describe the behavior of all types of waves. Radio waves from a satellite in orbit increase in frequency as the satellite moves closer to a receiver on Earth and drop in frequency as it moves away. Light waves, including those we

receive from stars millions of light years away, behave in the same way, though with light waves the change in frequency means a change in the color of the light. The application of Doppler's principle to starlight has helped astronomers explain a phenomenon they call the red shift, and it has led them to the awe-inspiring hypothesis that our universe is expanding, with stars and even galaxies moving away from each other at ever-increasing velocities.

Here's how Doppler's approaching and receding train whistles are used to posit a universe in motion. The color of light depends on its frequency. A prism will spread out white light (say from a star or galaxy) into a spectrum whose blue end is composed of high-frequency light and whose red end is composed of low-frequency light. When a light source is moving away from us, the frequencies are all shifted to lower ones—toward the red end of the spectrum—by the Doppler effect, just as the frequencies of a receding sound are shifted to lower ones. The greater the velocity of recession, the greater the "red shift."

The light from distant stars and galaxies shows an evident shift toward the red end of the spectrum. For instance, the light from the Nebula in Hydra, some 1.2 billion light-years distant, has been "red-shifted" halfway across the spectrum. This fact allows astronomers to calculate that the Nebula is moving

WITH DOPPLER

away from us at the tremendous speed of 60,000 kilometers per second, or almost one-fifth the speed of light. The further away a galaxy is, the greater its red shift, and therefore astronomers conclude that our entire universe is expanding.

Demonstrating Doppler

Though it would take sophisticated equipment to demonstrate the Doppler effect using rays of light, you'll be able to do a demonstration with sound waves, if you get yourself a small cylindrical whistle and a 2-meter length of rubber tubing (the ribbed kind would be best) into which the whistle will fit snugly. After inserting the whistle into one end of the tube, hold the other end to your lips and blow. Then, keeping the sound of the whistle steady, swing the whistle end of the tube in a large circle. As the whistle moves toward students standing in front of you, the frequency of the waves will increase and the whistle's pitch will rise. Conversely, as the whistle moves away from the listening students, the frequency will drop and the pitch will be lower. If some sound reflects off the walls of the classroom, your students may not be able to hear the Doppler effect. In this case, simply look for a place in the room where this interference does not occur and where the effect will be obvious. You may want to practice with the tubing and whistle before trying the demonstration in

class to make sure you can produce a steady and sustained tone—and to make sure the whistle won't fly out in the middle of the demonstration.

The physical laws that govern the universe are as mysterious and difficult to understand as the stars, but they're

also as close as a railroad crossing. Discussing the Doppler effect will give your students a glimpse of how physics can explain the everyday *and* the extraterrestrial, and it may start some of them thinking about the relationship between the two.

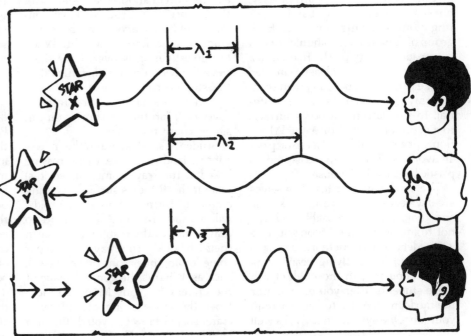

The Doppler Effect and the Red Shift

Star X is stationary with respect to the observer. Light with wavelength λ_1 travels to the observer. Star Y is moving away from the observer. Its light waves are stretched out so that their wavelength λ_2 is longer and their frequency is lower. Light from this star shifts to the red end of the spectrum. Star Z is moving toward the observer. The light waves are pushed closer together so that their wavelength λ_3 is shorter and their frequency is higher. Light from this star shifts to the blue end of the spectrum.

The Model Meteorite

by Jane W. Renaud

—Jane W. Renaud

How much do your students see when they look at pictures of the Moon's surface? By modeling meteor impact[1], students can gain firsthand information about the shape, size, and associated features of lunar craters.

Be forewarned, however, that this investigation may get a little messy; but it's a minor price to pay for getting your students involved. Each group of four students should make a newspaper "island" on the floor. Into a container with sides at least 5 cm deep (a cake pan, a shirt box, a garden seedling tray), pour some "sievable" material, to simulate the Moon's surface, to a depth of about 2 or 3 cm. Wheat flour, dry cement, or plaster of paris all work well and can be stored in plastic bags for later reuse.

The model meteorites themselves are dry beans: Navy beans work well, and they, too, are reusable. (This is not to imply that the Moon's surface is powdery or that meteorites behave like beans.) Since "flying beans" may not be popular with those who teach these students after you do, monitor the model meteorites to keep them in your lab. One system that works well

also reinforces another lab skill, use of the balance. Give students small containers of beans (about four beans per student) of a known weight. When the investigation is over, have them sieve the Moon material, return it to the storage bag, retrieve the beans, and weigh them in to assure "total recovery." If lab time is short, you can simply count the beans.

Students *stand* around the newspaper island and take turns tossing beans at the tray. Some (meteorites) will hit the "Moon's surface," causing craters to form. Others (meteoroids) will miss the Moon. Ask students to draw a map of the resulting surface in their lab books. You may want to put some leading questions on the board, such as: What is the general shape of the craters on your Moon surface? Does the angle of impact influence crater shape? How can you distinguish between early and late craters? What other features can you identify?

Dusting the model surface with a powder of contrasting color ("rotten stone" from a hardware store works well) makes rays and rim features much easier to distinguish. Students are usually surprised to find that cra-

ters are roughly circular, no matter what the meteorites' approach angle (as long as the beans are dropped from sufficient distance). And students quickly recognize the possibility of relative dating of impacts.

You can complete this activity in one time period. Follow it up with a discussion of actual Moon craters. Poster-size replicas of both sides of the Moon are available from the Superintendent of Documents, U.S. Government Printing Office, Washington, DC 20402. Almost any earth science book will have some illustrations of the Moon's surface features. Challenge your students to color code or otherwise identify at least three ages of craters—oldest, youngest, and in between. The area near *Mare Imbrium* (the largest of the lunar "seas") with the Copernicus ray system is a good area to examine for this, since at least five clearly separate time events are identifiable.

Happy landing! ∎

Jane W. Renaud is science coordinator at University Liggett School, 1045 Cook Rd., Grosse Pointe Woods, MI 48236.

---

[1]The original lab procedure from which this activity was developed was part of the Princeton Project, a study funded by the National Science Foundation.

Building a Telescope

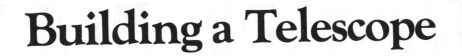

It's as easy as PVC.

by Chris F. Linas

The study of astronomy can seem as remote as the stars to beginning students. How can you help beginners focus in on this subject? Try giving them some pieces of PVC (polyvinylchloride) pipe—along with all the other parts they need to make their very own PVC telescopes.

I first learned about the PVC telescope at a physics institute conducted by Carl Wenning of Illinois State University. By modifying his original construction design, my colleagues (Fred Tarnow and Tony Brieler) and I were able to come up with a telescope perfect for the high school laboratory—it's inexpensive to build, requires no dangerous PVC cements or solvents, and calls for very little work with tools. So far, the telescope has passed inspection by my own general science and astronomy students and by a group of Illinois honors science teachers, all of whom built and used the telescopes in conjunction with star maps, phases-of-the-moon charts, and planet-locators.

Most of the parts for the PVC telescope are not sold in metric, so be prepared to shop in standard units. You'll have to buy two lenses for each telescope—an objective and an eyepiece. The PVC telescope's objective lens is an Edna Lite #522 optical copy lens. This 4.0-inch-long lens has a diameter of 3.0 in. and a casing 3.5 in. in diameter. It costs between $10 and $12, depending on the quantity of the purchase. The eyepiece, a 1-in. f/1.6, 8-mm movie projector lens, costs approximately $8.

You should be able to purchase these lenses through the American Science Center/Jerryco, Inc. (601 Linden Pl., Evanston, IL 60202), which specializes in surplus optical equipment. Edmund Scientific and your own school's audiovisual repair supplier may also provide these lenses at school discount prices.

Each telescope also requires the following parts: a section of PVC pipe 3 in. in diameter, 5 in. long, and 1/4 in. thick (about $.35 per student if purchased in 10-ft sections); a 3-by-1 1/2-in. PVC reducer (about $1.50 each); a 1 1/2-by-1 1/4-in. PVC reducer with an adjustable knob (also called a male compression fitting, about $1 each); a

—Art by Max-Winkler

Telescope Assembly

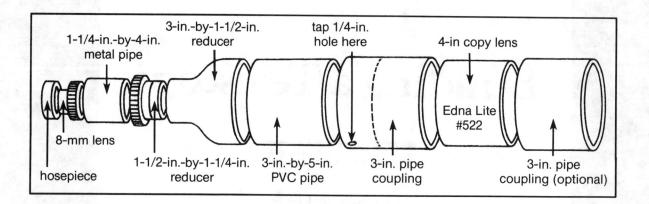

1-1/4-in.-by-4-in. metal pipe
3-in.-by-1-1/2-in. reducer
tap 1/4-in. hole here
4-in copy lens
8-mm lens
hosepiece
1-1/2-in.-by-1-1/4-in. reducer
3-in.-by-5-in. PVC pipe
3-in. pipe coupling
Edna Lite #522
3-in. pipe coupling (optional)

Insert the large end of the 8-mm lens into the 1 1/4-by-3/4-in. piece of hose until the end of the lens is flush with one end of the hose. (The small end of the lens should stick out slightly.) Place the entire piece of hose inside the metal pipe, large-lens-end first. The hose should fit tightly in the pipe—if it is loose, wrap a piece of tape around the hose and reinsert it.

Loosen the adjustable knob of the 1 1/2-by-1 1/4-in. reducer and insert the drain pipe into the 1 1/2-in. open-ing. Attach the 1 1/4-in. end of the reducer to the 1 1/2-in. end of the 3-by-1 1/2-in. reducer. You may have to use some force to get a tight fit.

Next, attach the 3-in. end of this reducer to one end of the PVC pipe and attach the pipe coupling to the other end of the pipe. Place the large copy lens in the other end of the coupling. Finally, if your teacher has supplied you with a second coupling, place that coupling over the other end of the lens. Your self-made telescope is ready for use.

section of chrome-plated copper drain-pipe or vacuum cleaner pipe 1 1/4 in. in diameter and 4 in. long (about $2 for a 12-in. section); a PVC coupling 3 1/2 in. in diameter and 3 in. long (about $1.25 each); and a section of automobile heater hose 1 1/4 in. in diameter and approximately 3/4 in. long (about $.10 per student if purchased by the foot). Buying a second 3-in. coupling to cover the copy lens at the end of the telescope is optional. Although this coupling is not essential to the operation of the telescope, it does enhance the telescope's appearance.

You should be able to buy all these parts at a hardware store, a plumbing supply store, or a lumber yard. Finally,

Chris F. Linas is the science supervisor of Indian Springs School District No. 109, 8001 S. 82nd Ave., Justice, IL 60458–1599. He is also an adjunct professor of computers in science education at Chicago State University and an adjunct professor of astronomy at the National College of Education, Chicago campus.

telescope construction also requires access to the following tools: a hacksaw (with spare blades) or a PVC pipe cutter, an electric drill with a 13/64-in. drill bit, and a 1/4-in. tapping bit.

Building the telescope

To begin, cut the PVC pipe that is 3 in. in diameter into 5-in.-long sections, cut the metal pipe into 4-in.-long sections, and cut the hose into 3/4-in.-long sections. Drill a 13/64-in. hole approximately 2 1/4 in. from one end of the coupling and then use the tapping bit to make a 1/4-in. tap at that site. (The tap hole provides a place to attach a tripod stand later.) If the students are doing the cutting and drilling, review safety precautions with them and make sure they are wearing protective eye gear before they start. Finally, have the students align and connect all the telescope pieces as shown in the box above. As long as the pieces are fitted together tightly,

no glue or screws are needed.

To focus on an object, turn the reducer knob in a counterclockwise direction until it becomes loose. Move the metal pipe in and out until the image becomes clear. Of course, the image will appear upside down, but in celestial observations position is relevant only to the observer, since no real "up" or "down" exists.

Carl Wenning has calculated that the PVC telescope's light-gathering power is about 57 times, and its resolving power about 15 times, that of the human eye. Its magnifying power is 9.5°. The apparent field of the eyepiece is about 50°; dividing the apparent field by the magnifying power gives the telescope's true field of view of about 5.3°.

The stars and planets will never seem too remote to students who have had a hand in the construction of their own observation equipment. So try sharing this interesting and rewarding activity with your class. ∎

Figure 1

Figure 2

Edible Stars

I require my Earth science students to learn the star positions of some of the more common constellations. I have found a fun way to review for their final test is to have an edible practice test.

I take each constellation and enlarge it on white butcher paper by using an overhead projector. I use small dots for the star positions on the paper. (For younger students, I would put connecting lines from star to star to help them better see the constellation outlines.) I then put each constellation on a separate sheet of paper, number the constellation sheets, and place them on my lab tables. I place a cookie over each dot to represent a star.

The students take turns table hopping" and writing down the names of the constellations. After they are finished identifying the constellations, we walk around the tables and discuss them. When this is done, the best part comes—eating the "stars."

Sharon Stroud
Janitell Jr. High School
Fountain, CO

Elliptical experimenting

Students can easily construct ellipses and discover their characteristics. While the time-honored technique of string-tack-cardboard works well for drawing small figures, large examples that the whole class can see may seem difficult to plot. However, I discovered that two small bathroom plungers work well as stable foci for drawing giant ellipses on the blackboard. (See Figure 1.)

To define an ellipse, construct an oval, and then draw the three lines shown in Figure 2. An ellipse is a plane closed curved line such that the sum of any two lines drawn from any point on the ellipse to the foci (two points inside the ellipse) is constant. This quantity equals the major axis, the maximum diameter of the ellipse. The size of an ellipse depends on the length of the major axis, whereas the shape of an ellipse depends on the distance between the two foci.

The trait students describe as "squashiness," a rather inelegant term, is more correctly referred to as the eccentricity of an ellipse, the ratio of the distance between the two foci and the major axis. Have your students draw an ellipse with an eccentricity of 0.5.

Notice that an ellipse becomes a circle when its eccentricity is zero and a parabola when its eccentricity equals one.
—*M.B.L.*

Bibliography

Bibliography

ASTRONOMY

Journal of College Science Teaching

Bisard, W. J. (1985). Practical Activities in Astronomy for Non-science Students. *Journal of College Science Teaching, 14(3)*, pp. 181–183.

Dukas, R. J., Jr., (1982). Charleshenge: An Archeo-Astronomy Exercise For The Elementary Lab. *Journal of College Science Teaching, 12(2)*, p. 99.

Page, T. (1983). Astronomy with the Space Telescope. *Journal of College Science Teaching, 12(4)*, pp. 258–261.

Reaves, G. (1985). Astronomy for Undergraduate Non-science Majors. *Journal of College Science Teaching, 14(3)*, pp. 158, 222–223.

Scott, R. L. & Watson, J. (1986). 'Science And Society' in the Planetarium. *Journal of College Science Teaching, 15(4)*, pp. 259–260.

Scott, R. L. (1985). Teaching in the Planetarium. *Journal of College Science Teaching, 14(4)*, pp. 248–250.

The Science Teacher

Berr, S. (1984). Solar Spectacular. *The Science Teacher, 51(4)*, pp. 20–27.

Chaisson, E. J. (1984). The Invisible Universe. *The Science Teacher, 51(2)*, pp. 18–23.

Chatel, R. K., Sr., (1983). Idea Bank: Cigar-box Spectroscope. *The Science Teacher, 50(7)*, pp. 62–63.

Cole, P. R. & Mallon, G. L. (1987). One Planetarium—To Go! *The Science Teacher, 54(3)*, pp. 25–27.

Davis, J. L. (1984). Idea Bank: Turn Your Classroom into a Planetarium. *The Science Teacher, 51(6)*, p. 63.

Derrick, M. (1986). Idea Bank: Space Science Ideas. *The Science Teacher, 53(5)*, p. 65.

De Zeeuw, T. A. (1991). Jupiter and its Moons. *The Science Teacher, 57(5)*, pp. 64-66.

Drake, D. (1991). Doppler Ball Effects. *The Science Teacher, 57(6)*, p. 62.

Ford, T. (1985). Idea Bank: Mission Through the Solar System. *The Science Teacher, 52(9)*, p. 45.

Glenn, W. H. & Cullen, B. T. (1985). Idea Bank: Eggs and the Equinoxes. *The Science Teacher, 52(1)*, p. 52.

Glenn, W. H. (1984). Halley's Comet Makes a Comeback. *The Science Teacher, 51(1)*, pp. 38–44, 49.

Glenn, W. H. (1985). Finding Comet Halley. *The Science Teacher, 52(5)*, pp. 64–67.

Glenn, W. H. (1987). The Modern Astronomy of Ancient Observatories. *The Science Teacher, 54(4)*, pp. 20–28.

Hammond, D. E. (1983). Racing Through Space. *The Science Teacher, 50(5)*, pp. 48–51.

Hoff, D. B. & Skinner, S. (1986). Idea Bank: Inventing Your Own Sky Myths. *The Science Teacher, 53(8)*, p. 49.

Iadevaia, D. G. (1984). Star Light, Star Bright. *The Science Teacher, 51(6)*, pp. 50–53.

Iadevaia, D. G. (1986). Solar Surface Phenomena. *The Science Teacher, 53(7)*, pp. 19–23.

Kelsey, L. (1985). Idea Bank: An Inexpensive Star Chart. *The Science Teacher, 52(9)*, p. 47.

Keown, D. (1982). Scaling Our Cosmic Prisons. *The Science Teacher, 49(1)*, pp. 52–54.

Kettner, J. (1989). Finding North and the Local Time by Using the Sun and a Clock. *The Science Teacher, 55(7)*, p. 76.

Lamb, W. G. (1982). Is the Moon Really Larger on the Horizon. *The Science Teacher, 49(2)*, pp. 46–48.

Lamb, W. G. (1983). Build-It-Yourself Astronomy. *The Science Teacher, 50(3)*, pp. 48–51.

Lamb, W. G. (1988). 35-mm Astrophotography. *The Science Teacher, 55(1)*, p. 104.

Leyden, M. B. (1984). The Elliptical Johannes Kepler. *The Science Teacher, 51(8)*, pp. 52–56.

Lay, J. (1984). Idea Bank: Classroom in Space. *The Science Teacher, 51(2)*, p. 54.

Lyden, M. B. (1985). Refracting on William Herschel. *The Science Teacher, 52(6)*, pp. 30–33.

Mallon, G. L. (1987). Cosmic Journeys. *The Science Teacher, 54(3)*, pp. 28–30.

McLellan, K. K. (1985). Idea Bank: Wave Motion. *The Science Teacher, 52(6)*, p. 52.

McLure, J. W. (1984). Polaris, Mark Kummerfledt's Star, and My Star. *The Science Teacher, 51(5)*, pp. 80–82.

McPherson, C. M. & West, D. C. (1988). A Stellar Syllabus for Physical Science. *The Science Teacher, 55(5)*, pp. 53–56.

Murphy, R. S. & Kwasnoski, J. B. (1983). Years, Days, and Solar Rays. *The Science Teacher, 50(7)*, pp. 40, 45.

Reed, G. (1986). Ancient Astronomers Along the Nile. *The Science Teacher, 53(6)*, pp. 59–62.

Renaud, J. W. (1983). The Model Meteorite. *The Science Teacher, 50(6)*, p. 24.

Reynolds, R. F. (1984). New Window on the Universe. *The Science Teacher, 51(7)*, pp. 45–48.

Reynolds, R. F. (1986). Take the NEWMAST Cure. *The Science Teacher, 53(3)*, pp. 32–33.

Ronca, L. B. (1983). "Old News" from the Moon. *The Science Teacher, 50(6)*, pp. 20–23.

Russo, R. (1982). Astronomy by Day. *The Science Teacher, 49(3)*, pp. 54–55.

Russo, R. (1983). Sun Tracking: More Daytime Astronomy. *The Science Teacher, 50(5)*, pp. 72–74.

Russo, R. (1988). Shoot the Stars—Focus on Earth's Rotation. *The Science Teacher, 55(2)*, pp. 25–26.

Quimby, D. J. (1984). How To Orbit the Earth. *The Science Teacher, 51(1)*, pp. 54–56.

Saroka, L. G. (1990). The Scientific Method at Face Value. *The Science Teacher, 57(3)*, pp. 45–48.

Scott, R. L. (1983). Photographing the Night Sky (Without a Telescope). *The Science Teacher, 50(8)*, pp. 20–25.

Sheets, M. N. (1986). Idea Bank: Seasonal Changes in Solar Intensity. *The Science Teacher, 53(1)*, pp. 176–177.

Smith, R. D. (1989). A Mechanical Device for Drawing Ellipses on a Chalkboard. *The Science Teacher, 55(6)*, p. 56.

Speitel, T. (1987). Teach Satellite Motion with Logo Simulation. *The Science Teacher, 54(2)*, pp. 40–41.

Thaw, R. F. (1983). Idea Bank: Phases of the Moon. *The Science Teacher, 50(7)*, p. 63.

Unknown, (1983). First Few Minutes. *The Science Teacher, 50(5)*, pp. 82–83.

Uslabar, K. (1988) 3-D Constellations. *The Science Teacher, 55(1)*, pp. 103–104.

U. S. Naval Observatory, (1987). Taking Note: Computer-driven Astronomy. *The Science Teacher, 54(2)*, pp. 55–56.

Victor, R. C. (1982). A Summer Late-Night Special: A Total Eclipse of the Moon. *The Science Teacher, 49(5)*, p. 29.

Watson, J., Jr., (1982). Celestial Sphere. *The Science Teacher, 49(5)*, p. 47.

Wong, O. (1986). Idea Bank: How Would You Send Someone to the Moon? *The Science Teacher, 53(8)*, p. 50.

Wong, O. K. (1986). Idea Bank: Construct a Model of the Expanding Universe. *The Science Teacher, 53(6)*, pp. 77–78.

Wong, O. K. (1987). Idea Bank: How Wide Is the Moon. *The Science Teacher, 54(7)*, p. 57.

Wong, O. K. (1988). Idea Bank: Proving that the Earth Is Spherical. *The Science Teacher, 55(5)*, pp. 65–66.

Science Scope

Ballinger, S. (1988). Make a Moon Calendar. *Science Scope, 11(5)*, pp. 20–21.

Davis, J. L. (1985). Once upon a Time When the Earth Was Flat. *Science Scope, 9(1)*, pp. 30–31.

Glasgow, D. (1988). Imaginary Stars. *Science Scope, 12(2)*, p. 37.

Hartman, J. (1990). Planning a Planetary Vacation. *Science Scope, 13(5)*, p. 26.

Hassard, J. (1990). Intergalactic and Oceanographic Missions. *Science Scope, 13(7)*, pp. 36–37.

Jividen, J. (1985). Potpourri: Active 'Star' Formations. *Science Scope, 9(2)*, p. 25.

Koballa, T. R., Jr., (1987). Why Don't Eclipses Occur Every Month? *Science Scope, 11(2)*, pp. 10–11.

Kupichek, J. (1991). Spacelink: Connect with NASA. *Science Scope, 14(8)*, p. 44.

Lewis, B. (1989). Build a Geodesic Dome. *Science Scope, 13(2)*, pp. 28–30.

Louviere, J. P. (1988). The Planetary Senate. *Science Scope, 12(3)*, pp. 10–11.

National Science Foundation (1991). Eyes on the Earth—The Frontiers of Remote Sensing. *Science Scope, 14(7)*, pp. 19–20.

Nickerson, J. (1989). A Basketball "World." *Science Scope, 12(5)*, p. 23.

Reynolds, J. (1987). Designing a Space Station. *Science Scope, 11(1)*, p. 24.

Riddle, B. (1990). Comet Watch–Comet Austin. *Science Scope, 13(7)*, p. 22.

Riddle, B. (1990). Scope on the Skies. *Science Scope, 14(1)*, pp. 38–39.

Riddle, B. (1990). Scope on the Skies: Tracking Mars. *Science Scope, 14(3)*, pp. 48–49.

Riddle, B. (1991). Scope on the Skies: Bright Starlight. *Science Scope, 14(4)*, pp. 24–25.

Riddle, B. (1991). Scope on the Skies: A Regal Mane. *Science Scope, 14(6)*, pp. 16–17.

Riddle, B. (1991). Scope on the Skies: Beauty ande Love. *Science Scope, 14(7)*, pp. 16–17.

Riddle, B. (1991). Scope on the Skies: A Spectacular Summer. *Science Scope, 14(8)*, pp. 41–42.

Ritchey, M. (1989). Trajectory to Mars. *Science Scope, 12(5)*, p. 23.

Rosene, D. (1990). Thirty Hours in Space—Sort of. *Science Scope, 14(3)*, pp. 16–20.

Signore, J. A. (1987). Computers as Standard Investigative Tools. *Science Scope, 10(5)*, pp. 10–11.

Starshine (1991). Star Clock. *Science Scope, 14(8)*, pp. 34–37.

Steinheimer, M. (1984). Webbing an Integrated Science Program. *Science Scope, 8(2)*, pp. 10–11.

Stroud, S. & Stroud, R. (1983). The Circumpolar 5 & the Heavenly G. *Science Scope, 7(1)*, pp. 10–11.

Stroud, S. (1983). Potpourri: Edible Stars. *Science Scope, 7(1)*, p. 9.

Sunal, D. W. (1984). Star Gazing. *Science Scope, 7(4)*, pp. 16–17.

Watson, N. T. & Scott, R. L. (1984). The Planetarium in Middle School Science. *Science Scope, 7(3)*, p. 15.

Wilson, S. (1987). Does the Sun Measure Up? *Science Scope, 10(3)*, p. 129.

Science & Children

Anthony, M. (1983). Moonstruck In Mansfield? *Science & Children, 20(7)*, pp. 10–11.

Bar, V. (1985). Following the Sun in Jerusalem. *Science & Children, 22(5)*, pp.16–18.

Bires, N. B. (1986). To Neptune and Back or Donna's Cosmic Adventure. *Science & Children, 23(5)*, pp. 20–24 & 58–62.

Cassen, J. (1984). Helpful Hints: Why the North Star Doesn't Move. *Science & Children, 21(7)*, p. 20.

Fehrenbach, C. R., Greer, F. S., & Daniel, T. B. (1986). LEA on the Moon. *Science & Children, 23(5)*, pp. 15–17.

Flowers, J. D. (1984). Evening Skies in a Shoebox. *Science & Children, 21(8)*, p. 55.

Geller, L. G. (1983). About Maps and the Moon— Learning from First Graders. *Science & Children, 20(7)*, pp. 4–5.

Harrington-Lueker, D. (1984). The November Science Calendar: A Classroom Weather Station. *Science & Children, 22(2)*, p. 27.

Hermann, C. (1987). The Perfect Mix: Science and Myth. *Science & Children, 24(8)*, pp. 26–27.

Hoff, D. B. (1983). Starry Archeology. *Science & Children, 20(5)*, pp. 26–27.

Hurd, M.D. (1991) Teach by the Light of the Moon. *Science & Children, 28(7)*, pp. 22-24

Joslin, P. H. (1985). As Smart as a Fencepost. *Science & Children, 22(5)*, pp. 13–15.

Katz, P. (1986). Hot Stuff. Science & Children, 23(5), p. 47.

Kepler, L. (1986). Time on Your Hands—How Long until Sundown? *Science & Children, 24(1)*, p. 35.

Kepler, L. (1988). Make Someone a Star. *Science & Children, 26(3)*, p. 52.

Lamb, W. G. (1984). Why Is The Sky Blue? *Science & Children, 21(4)*, pp. 101–102.

Leyden, M. B. (1984). For It Is Only a Paper Moon. *Science & Children, 21(7)*, pp. 18–20.

Leyden, M. B. (1985). Early Adolescence: Celestial Meetings—Synodic Periods. *Science & Children, 22(7)*, pp. 41–42.

Lightman, A. & Sadler, P. (1988). The Earth Is Round? Who Are You Kidding? *Science & Children, 25(5)*, pp. 24–26.

Loud, G. E. (1986). Off to the Moon. *Science & Children, 23(6)*, pp. 10–12.

Main, J. (1991). Blast Off! *Science & Children, 28(7)*, pp. 10–12.

Marames, K. (1986). Earth in Space: A Science Teach-In. *Science & Children, 24(3)*, pp. 36–38.

Martin, K. (1982). Helpful Hints: A Box Full of Stars. *Science & Children, 20(3)*, p. 8.

Muente, G. (1988). Helpful Hints: The Sun and the Golf Ball. *Science & Children, 25(7)*, p. 30.

Riddle, B. (1988). E.T., Call Nome . . . or Phoenix, or Burbank. *Science & Children, 25(4)*, pp. 30–33.

Sarko, T. (1983). Down-to-Earth Astronomy. *Science & Children, 20(4)*, p. 85.

Schatz, D. (1985). Orbiting with Halley's. *Science & Children, 23(3)*, pp.10–13.

Schroeder, R. (1990) Big Dipper Chicken with Gravy. *Science & Children, 28(1)*, p. 55.

Slavin, L. (1983). The Sky's the Limit. *Science & Children, 20(6)*, pp. 20–22.

Strickland, K. (1986). Fantasy. *Science & Children, 24(3)*, pp. 51–53.

Victor, R. C. (1982). May Evening Skies: A Study Guide. *Science & Children, 19(7)*, pp. 34–35.

Victor, R. C. (1982). Observing the Motions of the Planets. *Science & Children, 19(6)*, pp. 40–41.

Wahla, J. C. (1982). Using the Overhead to Stargaze. *Science & Children, 19(8)*, pp. 19–21.

Walker, O. H. (1984). Making Waves with Doppler. *Science & Children, 21(6)*, pp. 8–9.

West, R. (1988). Martians R Us. *Science & Children, 25(4)*, pp. 26–28.

GEOLOGY

Journal of College Science Teaching

Frichter, L. S. (1987). Teaching the Deep Structure of the Natural Sciences to Liberal Studies Students. *Journal of College Science Teaching, 17(1)*, pp. 11, 87.

Holmes, S., Lerche, I., Pantano, J., & Thompson, R. (1986). Politics, Garbage, and Fast-Food Franchises. *Journal of College Science Teaching, 16(1)*, pp. 121–122.

Mantei, E. J. (1986). Department-Generated Microcomputer Software. *Journal of College Science Teaching, 15(6)*, pp. 516–518.

McKenzie, G. D., Utgard, R. O., & Lisowski, M. (1986). The Importance of Field Trips: A Geological Example. *Journal of College Science Teaching, 16(1)*, pp. 17–20.

Moran, J. M. & Stieglitz, R. D. (1982). A Conceptual Framework for Teaching Glacial Geology. *Journal of College Science Teaching, 11(4)*, pp. 223–224.

Tourtellotte, S. W. (1987). Inexpensive Equipment for a Demonstration of Diastrophism. *Journal of College Science Teaching, 17(2)*, pp. 156–157.

The Science Teacher

Alexander, G. (1991). Sonar-Based Science. *The Science Teacher, 58,(5)*, p. 44-48.

Barry, M. A. (1985). Idea Bank: Radioactivity and Half Life. *The Science Teacher, 52(8)*, pp. 46–47.

Bissonette, T. E. (1983). Idea Bank: Crystals and Their Patterns. *The Science Teacher, 50(2)*, p. 54.

Brush, S. G. (1987). Cooling Spheres and Accumulating Lead: The History of Attempts to Date the Earth's Formation. *The Science Teacher, 54(9)*, pp. 29–34.

Cain, M. F. (1982). Idea Bank: Trashbasket Geology. *The Science Teacher, 49(9)*, p. 43.

Carnes, R. (1987). What is a Hundred Meters Long and a Hundred Million Years Old? *The Science Teacher, 54(2)*, pp. 20–21.

Carter, K. C. (1984). Idea Bank: Rock Porosity and Scientific Methodology. *The Science Teacher, 51(9)*, p. 49.

Chahrour, J. (1988). Idea Bank: Protecting Topographic Maps. *The Science Teacher, 55(2)*, p. 63.

Dodd, J. (1989). Paleontologists for a Day: A Simulated Fossil-find Activity. *The Science Teacher, 55(8)*, pp. 70-72.

Gartman, V. M. (1983). Idea Bank: Paleontology: Old and New. *The Science Teacher, 50(1)*, p. 66.

Glenn, W. H. (1983). Drifting: Continents on the Move. *The Science Teacher, 50(2)*, pp. 20–26.

Glenn, W. H. (1983). The Jigsaw Earth—Putting the Pieces Together. *The Science Teacher, 50(1)*, pp. 31–37.

Gulley, K. K. (1985). Idea Bank: Maps and Models. *The Science Teacher, 52(8)*, p. 46.

Harms, M. (1989). Sand and Soil Samples, *The Science Teacher, 55(8)*, p. 66.

Hicks, A. H. (1987). Idea Bank: Stratification of Water Temperature. *The Science Teacher, 54(1)*, p. 158.

Hirschfelt, J. (1988). Idea Bank: Teaching Geologic Time with a Computer. *The Science Teacher, 55(4)*, p. 46.

Hoekstra, B. (1982). The Groundwater Terrarium: Geology on a Small Scale. *The Science Teacher, 49(5)*, pp. 44–46.

Huckstep, C. (1985). Idea Bank: Finding and Determining Hard Water. *The Science Teacher, 52(9)*, p. 46.

Johnson, P. (1986). Cavebusters. *The Science Teacher, 53(2)*, pp. 20–22.

King, D. T., Jr. & Abbott-King, J. P. (1985). Field Lab on the Rocks. *The Science Teacher, 52(4)*, pp. 53–55.

Krockover, G. (1990). Putting Earth Science to the Test. *The Science Teacher, 57(3)*, pp. 30–33.

Kritsky, G. (1987). A New Day for Dinosaurs. *The Science Teacher, 54(1)*, pp. 28–31.

Lary, B. E. and Krockover, G. H. (1987). Maps, Plates, and Mount Saint Helens. *The Science Teacher, 54(5)*, pp. 59–61.

Leuenberger, T. (1986). Idea Bank: 1001 Uses. *The Science Teacher, 53(1)*, pp. 178–180.

Lewis, C. (1986). Idea Bank: Rock Exchange. *The Science Teacher, 53(4)*, p. 51.

Lockley, M. (1984). Dinosaur Tracking. *The Science Teacher, 51(1)*, pp. 18–24.

Lunetta, V. N., Wielert, J. S., & de Laeter, J. R. (1984). Science Around the Corner. *The Science Teacher, 51(1)*, pp. 50–53.

Maier, L. L. (1988). Idea Bank: Copper Hydroxide Stalactites and Stalagmites. *The Science Teacher, 55(1)*, p. 58.

Matthys, A. M. (1986). Idea Bank: Measuring the Age of Brick. *The Science Teacher, 53(3)*, p. 49.

Mattingly, R. L. (1987). The Dynamics of Flowing Water. *The Science Teacher, 54(9)*, pp. 23–27.

McKenzie, G. D. (1985). Tips for the Field. *The Science Teacher, 52(6)*, p. 37.

McLure, J. W. (1985). Free Tips for Geology Trips. *The Science Teacher, 52(7)*, pp. 41–43.

McMahan, R. & Parker, M. L. (1985). Idea Bank: The Cardboard Cave. *The Science Teacher, 52(2)*, pp. 58–59.

Metz, R. (1983). Rocking With Geology. *The Science Teacher, 50(4)*, pp. 24–25.

Metz, R. (1985). Idea Bank: TV Geology Games. *The Science Teacher, 52(7)*, pp. 46–47.

Metz, R. (1986). County Capture: A Geological Review Game. *The Science Teacher, 53(7)*, pp. 38–39.

Metz, R. (1987). Idea Bank: Card Games, Geology Style. *The Science Teacher, 54(7)*, pp. 59–60.

Metz, R. (1989). Geology Bingo. *The Science Teacher, 55(8)*, p. 73.

Pearman, A. (1986). Idea Bank: Geology Timeline. *The Science Teacher, 53(1)*, p. 173.

Ponnamperuma, C. (1983). Cosmochemistry: The Earliest Evolution. *The Science Teacher, 50(7)*, pp. 34–39.

Saccoman, D. (1989). Flipped Over Science. *The Science Teacher, 56(1)*, p. 76.

Scheintaub, H. (1987). Earth Elements by Analogy. *The Science Teacher, 54(4)*, pp. 29–30.

Schwoebel, G. (1990). When East Is East (or Is it West?). *The Science Teacher, 56(7)*, pp. 70–72.

Smith, A. E. (1987). Idea Bank: How Earthquakes Cause Faulting. *The Science Teacher, 54(1)*, p. 158.

Sproull, J. D. (1985). The Best on Earth. *The Science Teacher, 52(5)*, p. 61.

Texley, J. (1989). Radon: Reducible Risks, Rational Remedies. *The Science Teacher, 55(5)*, pp. 42–53.

Vernon, W. (1988). Chocolate Chip Petroleum. *The Science Teacher, 55(1)*, pp. 105–106.

Voltmer, R. K. & Paulson, R. L. (1983). Elevate Your Students with Topographic Maps. *The Science Teacher, 50(2)*, pp. 30–31.

White, B. (1986). Idea Bank: Can Plants Break Rocks? *The Science Teacher, 53(1)*, p. 175.

Wright, R. G. (1986). Get Close to Glaciers with Satellite Imagery. *The Science Teacher, 53(8)*, pp. 22–28.

Zuerner, F. (1985). Science and Sightseeing. *The Science Teacher, 52(5)*, p. 63.

Science Scope

Barba, R. H. (1985). Fossil Study Activities. *Science Scope, 8(4)*, pp. 26–27.

Bly, L. J. (1987). Teaching Topographic Maps. *Science Scope, 11(1)*, pp. 22–23.

Calhoun, M. J. (1987). Potpourri: Rocky History. *Science Scope, 11(2)*, p. 17.

Dalton, E. A. (1991). Pipelines and Headlines. *Science Scope, 12(3)*, pp. 24–26.

Davies, H. M. (1989) National Park Getaways. *Science Scope, 12(5)*, pp. 16–17.

De Golia, J. (1988). Expedition Yellowstone. *Science Scope, 11(6)*, pp. 28–29.

Fields, S. F. (1991). How We Use Our Precious Land. *Science Scope, 14(4)*, pp. 30–31.

Francis, L. (1991). Earthly Explorations. *Science Scope, 14(6)*, p. 11.

Gotsch, M. & Harris, S. (1988). Soil Lab: Groundwork for Earth Science. *Science Scope, 11(5)*, pp. 8–11.

Graika, T. (1989). Soil Erosion. *Science Scope, 13(1)*, p. 58.

Hall, J. W. (1985). Exploring Cincinnati's Rocky Past. *Science Scope, 8(3)*, pp. 48–49.

Hanson, S. (1985). Potpourri: Your Students Can Be Gems. *Science Scope, 9(2)*, p. 25.

Hartman, L. (1984). Potpourri: Rock Cake Recipe. *Science Scope, 8(1)*, p. 11.

Insert Poster: "Conservation and the Water Cycle." *Science Scope, 10(5)*, pp. 20-21.

Ireton, F. (1987). Panning for Gold. *Science Scope, 10(5)*, p. 35.

Kimmich, R. (1985). Using a Rock Key. *Science Scope, 8(4)*, pp. 9–11.

Koker, M. (1991). Investigating Groundwater. *Science Scope, 14,(8)*, p. 10-15.

Leyden, M. B. (1984). The Water Planet—An Optical Illusion. *Science Scope, 8(1)*, p. 8.

Leyden, M. B. (1986). Exploring Nature's Angles. *Science Scope, 9(3)*, p. 13.

Maier, L. L. (1988). Potpourri: Popping Half Life. *Science Scope, 11(5)*, p. 41.

Malone, V. (1984). Model of the Earth's Forces. *Science Scope, 7(4)*, p. 18.

Maute, J. & Renken, S. (1984). Rocking and Rolling Through Geology. *Science Scope, 7(4)*, pp. 12–13.

O'Neil, J. P. (1986). Using Landsat Photos to Teach Earth Science. *Science Scope, 9(3)*, pp. 10–11.

O'Neil, J. P., Blobner, W., & Enge, R. (1985). Those Mysterious Rocks & Minerals. *Science Scope, 9(1)*, pp. 12–13.

Palmer, A. C. (1987). Sandy Science. *Science Scope, 11(3)*, pp. 22–23.

Schneider, M. (1988). The Cave Adventure. *Science Scope, 11(4)*, pp. 6–8.

Selvig, L. (1988). Sedimentary, My Dear Watson. *Science Scope, 11(6)*, pp. 10–11.

Shellenberger, B. (1986). All About Rocks. *Science Scope, 10(2)*, pp. 12–13.

Texley, J. (1989) A Rational Approach to Radon. *Science Scope, 12(4)*, pp. 24–26.

Uslabar, K. (1986). Get Relief from Topographic Maps. *Science Scope, 10(2)*, p. 25.

Uslabar, K. (1987). Potpourri: Economics of Earth Science. *Science Scope, 10(3)*, p. 121.

Uslabar, K. (1988). Mapping Water Hardness. *Science Scope, 11(6)*, p. 9.

Villas, J. L. (1987). Giant Geologic Time Line. *Science Scope, 10(3)*, pp. 10–11.

Vandas, S. (1991). Water Use. *Science Scope, 14(8)*, pp. 27 & 43.

Wendel, W. C. (1985). The Planet Earth. *Science Scope, 9(2)*, p. 16 & Poster, Spinnaker Press, Inc.

Woerner, J. (1987). It's Your Fault. *Science Scope, 10(3)*, pp. 5–7.

Wright, L. & Dalton, E. A. (1987). All About Coal. *Science Scope, 11(3)*, p. 20.

Science & Children

Barba, R.H. (1990). What Does a Dinosaur Look Like? *Science & Children, 28(2)*, p. 37.

Barman, C. R. (1988). Soil of the Earth. *Science & Children, 25(4)*, pp. 38–39.

Brisard, W. (1988). Good Idea! *Science & Children, 26(1)*, pp. 20–21.

Coble, C. R. & McCall, G. K. (1986). Centering on Fossils and Dinosaurs. *Science & Children, 24(2)*, pp. 28–30, 50–51.

Crocker, B. & Shaw, E. L., Jr. (1985). Blueberry Trilobites and Peach Strata. *Science & Children, 22(8)*, pp. 10–11.

Croll, D. A. (1988). Helpful Hints: Did Dinosaurs Die of Cress. *Science & Children, 25(8)*, p. 29.

Denley, L. M. (1990). Primary Paleontology. *Science & Children, 28(2)*, pp. 30–32.

Fehrenbach, C. R., Green, F. S., and Barnes, B. (1989). Dinosuars Back in the Classroom. *Science & Children, 25(4)*, pp. 12–14.

Garver, J. B. (1983). Early Adolescence: Minerals and Formal Thought. *Science & Children, 21(1)*, pp. 54–56.

Hobby, E. J. (1982). Chocolate Chip Geology. *Science & Children, 20(3)*, p. 31.

James, C. (1991). Investigating Energy. *Science & Children, 28(7)*, pp. 18–19.

Kepler, L. (1988). Date with Science: Stalagmites and Stalactites. *Science & Children, 25(8)*, p. 6.

Klein, S. E. (1982). Making and Mining a Mountain. *Science & Children, 19(6)*, pp. 7–8.

Learner, R. J. (1983). Dinosaur Songs. *Science & Children, 20(5)*, pp. 14–18.

Lightman, A. & Sadler, P. (1988). The Earth Is Round? Who Are You Kidding? *Science & Children, 25(5)*, pp. 24–26.

McBryde, B. E., & Brown, F. W. (1985). When Water Cools Down. *Science & Children, 22(4)*, pp. 23–24.

McQuade, F. (1986). Interdisciplinary Contours: Art, Earth Science, & LOGO. *Science & Children, 24(1)*, pp.25–27, 85.

National Energy Foundation (1989). From Mountains to Metals. *Science & Children, 27(3)*, p. 41.

Nelson, D. J. (1985). Dirt Cheap and All Around Us. *Science & Children, 22(6)*, pp. 10–12.

Rice, D. & Corley, B. (1987). Talking Rocks. *Science & Children, 25(2)*, pp. 14–17.

Robinson, N. P. (1988). Helpful Hints: Sand and Things. *Science & Children, 25(7)*, p. 27.

Schamp, H. (1988). And the Slate Shale Be Cleaved. *Science & Children, 25(5)*, pp. 34–35.

Szymanski, C. (1984). Inferring about Dinosaurs. *Science & Children, 22(3)*, p.12–15.

Van De Walle, C. (1988). Water Works. *Science & Children, 25(7)*, pp. 15–17.

Vandas, S. (1991). Water: The Resource that Gets Used and Used and Used for Everything. *Science & Children, 28,(8)*, p. 8-9.

Wong, O. (1991) Prehistoric Perspectives. *Science & Children, 28(8)*, p. 24.

METEOROLOGY

Journal of College Science Teaching

Kauffman, G. (1985). A Crushing Experience. *Journal of College Science Teaching, 14(4)*, p. 364.

The Science Teacher

Bybee, R. (1984). Acid Rain: What's the Forecast? *The Science Teacher, 51(3)*, pp. 36–40, 45–47.

Bybee, R., Hibbs, M., & Johnson, E. (1984). The Acid Rain Debate. *The Science Teacher, 51(4)*, pp. 50–55.

Coolidge, B. W. (1987). Our School Gave Acid Rain a Hearing. *The Science Teacher, 54(3)*, pp. 31–33.

Knorr, T. P. (1984). Weather or Not to Teach Junior High Meteorology. *The Science Teacher, 51(5)*, pp. 62–64.

LaHart, D. E. & Gleman, S. M. (1982). Tapping the Sun: Blueprint for an Experimental Water Heater. *The Science Teacher, 49(1)*, pp. 28–33.

Lathrop, D. (1989). Coriolis Effects. *The Science Teacher, 56(4)*, p. 53.

Martin, H. E. (1987). Could *You* Build a Satellite Tracking Station? *The Science Teacher, 54(1)*, pp. 14–17.

McGregor, K. M. (1986). A Tornado in Your Classroom. *The Science Teacher, 53(4)*, pp. 16–19.

McLellan, K. K. (1985). Idea Bank: Wave Motion. *The Science Teacher, 52(6)*, p. 52.

Mogil, H. M. (1983). Weather and the W.C. *The Science Teacher, 50(1)*, pp. 28–30.

Murphy, R. S. & Kwasnoski, J. B. (1983). Years, Days, and Solar Rays. *The Science Teacher, 50(7)*, pp. 40, 45.

Olson, G. (1989). Making a Cloud in a Cheap Bottle. *The Science Teacher, 56(1)*, p .76.

Ojala, C. F. & Ojala, E. J. (1987). Airborne Particles. *The Science Teacher, 54(6)*, pp. 41–42.

Palmer, A. C. (1986). What Tree Rings Tell. *The Science Teacher, 53(6)*, pp. 70–73.

Rasmusson, E. M. (1984). Global Weather's Problem Child—El Nino. *The Science Teacher, 51(7)*, pp. 24–28.

Sheets, M. N. (1986). Idea Bank: Seasonal Changes in Solar Intensity. *The Science Teacher, 53(1)*, pp. 176–177.

Smith, D. R. & Krockover, G. H. (1988). Atmospheric Science: It's More Than Meteorology. *The Science Teacher, 55(1)*, pp. 36–39.

Sonnier, I. L. (1984). Idea Bank: Fog and Cloud Chamber. *The Science Teacher, 51(2)*, p. 53.

Summers, R. J. (1982). Satellite Weather Watch. *The Science Teacher, 49(4)*, pp. 43–46.

Science Scope

Adamopoulos, A., Dennis, C., Iamonte, B., Messersmith, J., and Osner, P., (1988). Weather Maps—Out of the Fog and Into the Clear Air. *Science Scope, 12(3)*, pp. 42–44.

Barrow, L. H. (1984). A pH Model for Studying Acid Rain. *Science Scope, 7(3)*, p. 9.

Insert Poster: "Conservation and the Water Cycle." *Science Scope, 10(5)*.

Jarcho, I. S. (1984). An Introduction to Acid Rain. *Science Scope, 7(3)*, pp. 6–8.

Malone, V. & Hutto, G. (1985). Challenge Your Students with Air Pressure Experiments. *Science Scope, 8(4)*, pp. 28–29.

Maute, J. (1986). Cloud Creativity. *Science Scope, 9(3)*, p. 9.

May, D. (1984). Experimenting with Air Pressure. *Science Scope, 8(1)*, p. 14.

Meier, B. (1988). Compute the Weather in Your Classroom. *Science Scope, 12(3)*, pp. 20–21.

Mogil, H. M. (1986). Headlining Meteorology. *Science Scope, 9(3)*, pp. 6–8.

Roberts, F. (1985). Using Weather Projects to Make Science Relevant. *Science Scope, 8(4)*, p. 33.

Rosene, D. (1987). Potpourri: It's Snow Joke. *Science Scope, 10(3)*, p. 121.

Saunders, W. L. (1985). The Expansion of Water Upon Freezing. *Science Scope, 9(2)*, p. 14.

Selnes, M. (1986). Science Startlers. *Science Scope, 10(1)*, pp. 5–7.

Selnes, M. D. (1983). Potpourri: Tumbler and Handkerchief. *Science Scope, 7(2)*, p. 9.

Uslabar, K. (1987). Possible Pathways Along the Water Cycle. *Science Scope, 10(5)*, pp. 20–21.

Wesner, T. (1985). A School Weather Station. *Science Scope, 8(3)*, pp. 52–53.

Science & Children

Acid Rain Foundation. (1987). Acid Rain Units from ARF. *Science & Children, 24(8)*, p. 28. "Acid Rain Poster", Between pp. 28–37, Acid Rain Foundation, Inc., St. Paul, MN.

Brown, P. S. (1982). Patterns in the Sky. *Science & Children, 19(4)*, pp. 18–19.

Brown, P. S. (1983). A Thunderstorm Strikes. *Science & Children, 20(8)*, pp. 18–20.

Hanif, M. (1984). Acid Rain. *Science & Children, 22(3)*, pp. 19–23.

Harbster, D. A. (1984). Balloons and Blow Battles. *Science & Children, 22(3)*, p. 6–8.

Harbster, D. A. (1988). Air Apparent. *Science & Children, 25(8)*, pp. 12–14.

Harrington-Lueker, D. (1984). The November Science Calendar: A Classroom Weather Station. *Science & Children, 22(2)*, p. 27.

Kepler, L. (1986). Thunder and Lightning. *Science & Children, 24(1)*, p. 41.

Kepler, L. (1988). How Clean Is Your Snow? *Science & Children, 25(3)*, p. 54.

Kepler, L. (1988). Date with Science: Making Clouds. *Science & Children, 25(6)*, p. 8.

Kibbey, B. R. (1987). Twister in a Bottle. *Science & Children, 24(8)*, p. 13.

Klooster, D. J. (1987). Understanding pH Through Acid Rain. *Science & Children, 24(8)*, p. 37.

Lamb, W. G. (1984). Why Is the Sky Blue? *Science & Children, 21(4)*, pp. 101–102.

Marames, K. (1986). Earth in Space: A Science Teach-in. *Science & Children, 24(3)*, pp. 36–38.

McFee, E. (1985). A Jugful Of Science. *22(8)Science & Children*, pp. 24 & 41.

McKee, J. (1988). Helpful Hints: Rainbows. *Science & Children, 25(7)*, p. 29.

Mogil, H. M. (1984). Sharpening Your Weather Eye. *Science & Children, 21(5)*, pp. 14–18.

Mogil, H. M. (1984). Winter Storm Watch. *Science & Children, 21(6)*, pp. 10–14.

Mogil, H. M. (1986). Are You Flying Out to 'Frisco? *Science & Children, 23(6)*, pp. 38–39.

Mogil, H. M. & Collins, H. T. (1989). Geography and Weather—Hurricane. *Science & Children, 27(1)*, pp. 37–44.

Mogil, H.M. & Collins, H. T. (1989). Geograpy and Weather— Seasons. *Science & Children, 27(3)*, pp. 33–40.

Mogil, H. M. & Collins, H. T. (1990). Geography and Weather—Mountain Meteorology, *Science & Children, 27(5)*, pp. 29–33.

Mogil, H. M. & Levine, B. G. (1990). Geography and Weather—Weather Mapping. *Science & Children, 28(3)*, pp. 31–38.

Mogil, H. M., Beller-Simms, N., & Levine, B. G. (1991) Geography and Weather—Deserts. *Science & Children, 28(7)*, pp. 22-24.

Padilla, M. (1983). Early Adolescence: Snowy Science. *Science & Children, 20(4)*, pp. 98–99.

Pampe, W. R. (1985). Mirages. *Science & Children, 23(3)*, pp. 21–23.

Pampe, W. R. (1986). A Hurricane! *Science & Children, 24(1)*, pp. 16–19, 55.

Pippel, M. (1991). Invasion of the Snow Monsters. *Science & Children, 28(4)*, pp. 18–19.

Plesko, F. (1984). Balloon Talk—Acid Rain. *Science & Children, 22(3)*, pp. 44–45.

Probasco, G. (1982). In the Schools: Balloons Away. *Science & Children, 20(2)*, pp. 20–21.

Schafer, L. (1986). Permanent Snowflakes. *Science & Children, 24(3)*, pp. 11–13.

Shaw, J. M. & Owens, L. L. (1987). Weather—Beyond Observation. *Science & Children, 25(3)*, pp. 27–28.

Stepans, J. & Kuehn, C. (1985). Children's Conception of Weather. *Science & Children, 23(1)*, pp. 44–47.

Wilson, L. H. (1989). Windsocks. *Science & Children, 25(5)*, p. 53.

OCEANOGRAPHY

Journal of College Science Teaching,

Bokuniewicz, H. & Cerrato, R. (1986). Long Island Sound: Science and Use. *Journal of College Science Teaching, 16(1)*, pp. 40–42.

Myers, R. L. (1983). An On-Land Approach to Teaching the Realities of Oceanography. *Journal of College Science Teaching, 13(1)*, pp. 26–27.

Saveland, R. N. & Stoner, A. W. (1985). Mission Possible: The Sea Semester Program. *Journal of College Science Teaching, 14(6)*, pp. 484–488.

Wallace, W. J. (1982). A Water Course. *Journal of College Science Teaching, 11(5)*, pp. 299–300.

The Science Teacher

Alexander, G. (1991). Sonar-based Science. *The Science Teacher, 57(9)*, p. 58.

Gross, R. (1985). Idea Bank: Why Not a Hydrometer? *The Science Teacher, 52(8)*, p.48.

Jesberg, R. O. & Dowden, E. (1986). Microchip Measuring. *The Science Teacher, 53(7)*, pp. 34–37.

Klemm, E. B. (1991) A Rapid Way to Test Liquid Density. *The Science Teacher, 57(6)*, p .58.

Science Scope

Lee, J. (1987). Potpourri: The Ice Cream Lab. *Science Scope, 10(3)*, p. 9.

Reynolds, K. E. (1987) Reynolds' Rap: Draw a Consequence Map. *Science Scope, 10(5)*, p. 22.

Selvig, L. (1986). Mapping the Ocean Floor. *Science Scope, 10(1)*, p. 30.

Stewart, L. A. (1988). Hydrometer Lab. *Science Scope, 11(4)*, pp. 18–19.

Stone, M. (1991). The Unseen Bottom. *Science Scope, 14(4)*, pp. 32–35.

Wiley, M. (1984). Investigating The Ocean Floor from an Inland School. *Science Scope, 8(2)*, pp. 16–17.

Science & Children

Butler, V. R. & Roach, E. M. (1986). Gone with the Wind. *Science & Children, 24(2)*, p. 37.

Collins, H. T. & Mogil, H. M. (1990). Geography and Weather—Oceans. *Science & Children, 27(7)*, pp. 17–24.

Landry, A. & Janke, D. (1987). In the Schools: "O Christmas Tree." *Science & Children, 25(3)*, pp. 39–40.

Leachworth, M. D. (1991). Cool Stuff. *Science & Children, 28(8)*, p. 24.

Pahl, J. (1988). In the Schools: River at Gunpowder. *Science & Children, 25(7)*, pp. 36–37.